Glimpses of Creatures in Their Physical Worlds

Glimpses of Creatures in Their Physical Worlds

Steven Vogel

PRINCETON UNIVERSITY PRESS

PRINCETON AND OXFORD

Copyright © 2009 by Princeton University Press
Published by Princeton University Press, 41 William Street, Princeton, New Jersey 08540
In the United Kingdom: Princeton University Press, 6 Oxford Street,
Woodstock, Oxfordshire OX20 1TW

All Rights Reserved

Library of Congress Cataloging-in-Publication Data

Vogel, Steven, 1940–
    Glimpses of creatures in their physical worlds / Steven Vogel.
        p. cm.
    Includes bibliographical references and index.
    ISBN 978-0-691-13806-0 (hardback : alk. paper) — ISBN 978-0-691-13807-7 (pbk. : alk.
paper) 1. Biophysics.   I. Title.
    QH505.V625 2009
    571.4—dc22

                                                    2008049338

British Library Cataloging-in-Publication Data is available

This book has been composed in sabon

Printed on acid-free paper. ∞
nathist.princeton.edu

Printed in the United States of America

10 9 8 7 6 5 4 3 2 1

Contents

Preface

THE DOZEN ESSAYS herein look at bits of biology, bits that reflect the physical world in which organisms find themselves. Evolution can do wonders, but it cannot escape its earthy context—a certain temperature range, a particular gravitational acceleration, the physical properties of air and water, and so forth. Nor can it tamper with mathematics. Thus the design of organisms—the level of organization at which natural selection acts most directly as well as the focus here—must reflect that physical context. The baseline it provides both imposes constraints and affords opportunities, the co-stars in what follows.

I take what other biologists might find an unfamiliar approach—one, by the way, that I've found productive enough to recommend. Instead of asking about the physical science behind a specific biological system, I'll consider an aspect of the physical world and ask what organisms, any organisms, make of it, both how they might capitalize on it and how they might be limited by it. If this approach to science were a dart game, I'd be thrown out—for throwing darts at a wall first and only subsequently painting targets around the points of impact.

In their original incarnation, the essays appeared every three months in successive regular issues of the *Journal of Biosciences*, published by the Indian Academy of Sciences, in Bangalore. They've been slightly expanded (especially the first, which was hurriedly done), updated just a little, and subjected to some irresistible tinkering, tampering, and addition of both verbal and pictorial illustrations. They're thoroughly self-indulgent, from choice of topics and level to outlook, as they're neither reviews in quite the conventional sense nor journalistic pieces on the most recent "break-throughs." Rather, they reflect my personal perspectives, with a bias toward providing exposure for ideas that I think might be worth looking into about now.

Not only do I not claim any kind of proprietary rights to anything here, at least beyond the ordinary imperatives of the copyright on words, I urge that anyone whose interest might be piqued take the ball and run with it. If that happens, I'll do all that I can to minimize the reinvention of any wheels in return for satisfying my curiosity—no trivial reward as the experimental scientist superannuates. Perhaps I should put the matter in more general terms.

When I began to do science, nearly fifty years ago, I wondered first whether and then where I'd get ideas worth pursuing. Now, on the cusp of retirement, I wonder what I'm going to do with my accumulated head—and notebooks—full of questions. Maybe we need something like a patent expiration date—if one does nothing with a hypothesis for some number of years, it should somehow revert to the public domain. At this point I mean to function as a facilitator, and I mean to do it not just for any issues raised here, but as best I can for any others in the general area of biomechanics. Engineers and physical scientists find entry into the jargon-ridden biological literature challenging, while biologists as often don't know where to turn to make sense of relevant physical issues. I can at least claim long experience at the interface. In particular, as a classical biologist by origin, I think I know (and still carry as gaps and scars) the difficulties a biologist faces. Underlying these essays lies the tacit argument that one can do a surprisingly wide range of things in this sort of physical biology with what might seem to be a laughably inadequate formal background.

A few prefatory words about topics. A main objective here is to make the case through examples that addressing a far-from-insignificant range of biologically significant questions requires physical rather than chemical or genomic reduction. For few of the issues here can any kind of chemo-molecular explanation be imagined. The choice of topics, again, reflects the author's idiosyncrasies and personal history—I make no claim that the largely mechanical and thermal underpinnings exhaust the relevance of the physical sciences. Optics and radiation physics, acoustics, and other areas certainly play analogous roles, but I'm ill-suited to play Boswell for much beyond the present domain.

Still, the essays don't hew to the normal biomechanical canon—to use a biomechanical metaphor. Moving (and not moving) heat and keeping water in its liquid phase extend the present domain into thermal physics (or heat transfer engineering) and physical chemistry. On some days I see that as partial expiation for my own role as an author in setting a biomechanical canon. What needs emphasis rather than any apology is the continuity between these areas and our more normal concerns—should convective heat transfer be regarded as thermal or fluid mechanical engineering, or perhaps part of thermoregulatory physiology? The choice, it would seem, hinges on nothing more than experience and convenience. At least in the present case, candor compels admission, as well, of a fair dose of the irrationality that underlies more of our decisions than most of us comfortably admit—in particular my early fascination with the comparative physiology of figures such as Per Scholander and Knut Schmidt-Nielsen. What matters more is that we not lose sight of the basic triviality of such decisions about disciplinary boundaries. However the specifics might be specified, we're still looking at aspects of life's physical context.

I've stayed with the level of assumed background and quantification of my previous book, the textbook *Comparative Biomechanics* (Princeton University Press, 2003)—that of an ordinary biologist, with a bit of terminological help for people not close to our jargon-afflicted world. I hope that help proves adequate for a peculiar present demand. With the message that a given physical phenomenon may have relevance in a wide diversity of biological contexts comes a lot of jumping around from an aspect of the life of one organism in one habitat to a different aspect of another organism in another habitat.

A note to the reader coming from biology. The traditional academic organization puts physics departments in our own college but assigns mechanical engineering to another. So our exposure to physical science comes mainly from physicists, as do most of our other academic interactions. But beyond the introductory level—the first physics course often required of biologists—engineering books and the engineers themselves seem to be distinctly more useful than is what one encounters in most physics departments and libraries. Not that the physicists aren't fascinating folks—but both the questions they commonly ask and their accustomed level of analytical rigor and abstraction tend to put some distance between them and our concerns. Most of the simple formulas here can be more easily found in engineering books and followed up by recourse to the engineers themselves.

A note to the reader coming from physics or engineering. Biology, no less than the physical sciences, seeks broadly applicable rules, common patterns of organizations, and order beneath the perceptually chaotic world. But biology enjoys a strange organizing principle, evolution by natural selection, barely hinted at elsewhere in science. Both evolution and the sheer diversity of life put especially bad bumps in that search. Perhaps its special difficulty underlies the gradual estrangement of biology from the more obviously successful physics of the post-Newtonian era and the awkwardness of its reintegration into the larger world of science in the twentieth century. The tidy formulas of Newtonian physics work even less well for our distinctive and untidy subject than they do for, say, practicing engineers.

And a note for both disciplines. The original intent of the essays was to show how much of the biological impact of the physical world could be discussed with only the most minimal mathematics, and I've kept the resulting level of exposition. So the biologist or aspiring biologist should encounter few problems, and engineers and physical scientists will be (one hopes) amused and (one also hopes) indulgent. To all I declare emphatically that more mathematics and quantitative rigor will take one further and that these essays make no pretense of being other than broad-brush pictures.

Finally, for readers or downloaders of the original essays, what has been altered (besides minor items of style) in this collected version? Chapter 1, about the Péclet number, has been the most extensively extended, mainly because it was done in inadvisable haste the first time. Chapters 2 and 3, on ballistics, gained only a few new ideas but a lot more supporting data. While at it, proper references for the original articles are as follows.

Living in a physical world. I. Two ways to move material. *J. Biosci.* **29**: 391–397 (2004).
Living in a physical world. II. The bio-ballistics of small projectiles. *J. Biosci.* **30**: 167–175 (2005).
Living in a physical world. III. Getting up to speed. *J. Biosci.* **30**: 303–312 (2005).
Living in a physical world. IV. Moving heat around. *J. Biosci.* **30**: 449–460 (2005).
Living in a physical world. V. Maintaining temperature. *J. Biosci.* **30**: 581–590 (2005).
Living in a physical world. VI. Gravity and life in the air. *J. Biosci.* **31**: 13–21 (2006).
Living in a physical world. VII. Gravity and life on the ground. *J. Biosci.* **31**: 201–214 (2006).
Living in a physical world. VIII. Gravity and life in water. *J. Biosci.* **31**: 309–322 (2006).
Living in a physical world. IX. Maintaining liquid water. *J. Biosci.* **31**: 525–536 (2006).
Living in a physical world. X. Pumping fluids through conduits. *J. Biosci.* **32**: 207–222 (2007).
Living in a physical world. XI. To twist or bend when stressed. *J. Biosci.* **32**: 643–655 (2007).
Living in a physical world. XII. Keeping up upward and down downward. *J. Biosci.* **32**: 1067–1081 (2007).

While I have received help from a large number of people, one person played a special role. During the fall of 2004, the editor of the *Journal of Biosciences*, Vidyanand Nanjundiah, invited me to do this series. It was a most flattering expression of confidence, more than a bit daunting when I think about it that way. He gave me exceptional leeway in scope, approach, and level, not even suggesting a specific model for what he might have had in mind. And he both provided his own helpful guidance and arranged for well-informed outside reviews of each piece I submitted. I hope that Vidya feels, in retrospect, that he exercised good judgment. For better or worse, it is quite unlikely that I would otherwise have written what you find here.

Libraries play a critical role in assembling this kind of material. But all too often one depends on people to get some sense of what material is available, where it can be found, and what it means—and, most especially, what has not been done, what can't reasonably be done, and where one is barking up the wrong tree to begin with. I'm indebted to a host of librarians, beginning with Teddy Gray and Sarah Hodkinson at Duke. At the same time, I'd like to register a protest at the spreading practice of putting books in facilities where I can't work the shelves (pejoratively known as "browsing"). It eliminates those wonderful "aha" moments when one finds something central that one didn't know existed, besides makes a mockery of the notion of subject cataloging.

My computational savvy reflects both my age and long-term lack of an adequate urge for self-improvement. The computations that underlie these essays used a delightfully antique program called "Chipmunk Basic." For statistical tests—regressions and correlations—I turned to the Vassar College Statistical Computation Website, a product of Emeritus Professor Richard Lowry. Where not specifically attributed, photographs are my own efforts.

I made use of a variety of facilities while putting these essays together. Several were mainly written when I was Visiting Scholar at the Darling Marine Center of the University of Maine; for arranging that visit I am grateful to Sara Lindsay and to Kevin Eckelbarger, its director.

I'm indebted to the large number of people whom I badgered and cajoled in the effort to deal with a lot of material with which I often was laughingly unfamiliar. Those whose names I can recall include Dave Alexander, McNeill Alexander, Lewis Anderson,* Chris Barnhart, John Bush, Young-Hui Chang, Steve Churchill, Jonathan Cox, Hugh Crenshaw, Tom Daniel, Mike Dickison, Olaf Ellers, Shelley Etnier, Frank Fish, Morris Flynn, Bob Full, Mory Gharib, Tim Griffin, Ron Grunwald, Melina Hale, John Havel, Matthew Healy, Mary Henderson, Rob Jackson, Sönke Johnsen, Pete Jumars, Dwight Kincaid, Peter Klopfer, Dan Livingstone, Kate Loudon, Jeff Lucas, Paul Manos, Alphonse Masi, Larry Mayer, Molly McMullen, John Mercer, Laura Miller, Anne Moore, Fred Nijhout, Steve Nowicki, Kevin Padian, Robert Page,* Tim Pedley, Chuck Pell, Jane Philpott,* Sarah Prather, Marney Pratt, Jim Price, Andy Rapoff, Mike Reedy, Howard Reisner, Steven Rice, Andy Ruina, Knut Schmidt-Nielsen,* Kalman Schulgasser, Rick Searles, Kathleen Smith, Jake Socha, John Sperry, Nancy Stamp, Roman Stocker, Don Stone, Mami Taniuchi, Frances Trail, Lloyd Trefethen,* Vance Tucker, Scott Turner, Betty Twarog, Melvin Tyree, Janice Voltzow, Peter Wainwright, H K Wallace,* Martha Weiss, Will Wilson, Jeanette Wyneken, and Amy

Zanne. Those whose names I've marked with asterisks may be deceased, but they are not forgotten.

Steven Vogel
Durham, North Carolina

Glimpses of Creatures in Their Physical Worlds

CHAPTER 1

Two Ways to Move Material

INTRODUCING A VARIABLE

"No man is an island, entire of itself," said the English poet John Donne. Nor is any other organism, cell, tissue, or organ. We're open systems, continuously exchanging material with our surroundings as our parts do with their surroundings. In all of these exchanges, one physical process inevitably participates. In that process, diffusion, thermal agitation, and place-to-place concentration differences combine to produce net movements of molecules. On almost any biologically relevant scale, it can be described by exceedingly precise statistical statements, formulas that take advantage of the enormous numbers of individual entities moving around. Since it incurs no metabolic expenditure, it's at once dependable and free.

But except over microscopic distances, diffusion proceeds at a glacial pace. For most relevant geometries, doubling distance drops the rate of transport per unit time by a factor, not of two, but of four. Diffusive transport that would take a millisecond to cover a micrometer would require no less than a thousand seconds (17 minutes) to cover a millimeter and all of a billion (a thousand million seconds or 3 years) for a meter. Diffusion coefficient, the analog of conventional speed, has dimensions of length squared per time rather than length per time—it's not a rate in the ordinary sense.

Some organisms rely exclusively on diffusion to move material internally and to transfer it to and from their surroundings. Unsurprisingly, they're either very small or very thin or (as in many coelenterates and macroalgae) bulked up with metabolically inert cores. Diffusion coefficients in air run about 10,000 times higher than in water, which translates into a hundred-fold (the square root of 10,000) distance advantage. So under equivalent circumstances, those living in air or transferring gases (as do many arthropods) can get somewhat larger—perhaps one hundred-fold—but then still face that daunting size-dependence of diffusion. In response, one might say, macroscopic organisms inevitably augment diffusion with an additional physical agency, variously termed convection, advection, or just bulk flow, in any case fluid flow en masse. Circulatory systems as conventionally recognized represent only one version of this ubiquitous fix.

Indeed, the size scale at which life switches from reliance solely on diffusive exchange to convection supplementation—very roughly 10 micrometers—corresponds, roughly, to the switch from cellular to multicellular organization. While being essentially one- or two-dimensional does permit macroscopic size, it comes with obvious limitations. And while many plant cells, about which more shortly, do get comparatively large, they quietly practice intracellular bulk flow.

Solari et al. (2006) explore this transition point, using as material flagellated colonial green algae, mainly *Volvox*. In this genus, active cells populate the periphery of spherical colonies around 0.5 millimeters in diameter. The daughter colonies within (as in plate 1.1) depend on coordinated beating of the parental flagella on the outside of the colony to create enough external flow for adequate exchange of metabolites and wastes. Even at this relatively small size, flow plays an important role—in effect they have circulatory systems located around their external surfaces. Deflagellating colonies lowers photosynthetic productivity; providing forced external fluid motion (a bubbler in the suspension) restores normality.

One might expect good design to balance the two physical processes. Excessive reliance on diffusion would limit size, slow the pace of life, or require excessively surface-rich geometries. Excessive reliance on flow would impose an unnecessary cost of pumping (chapter 10) or require an unnecessarily large fraction of body volume for pipes, pumps, and fluid. So for biological systems a default ratio of convective transport to diffusive transport should be around one. As it happens, the chemical engineers provide us with just such a ratio. This so-called Péclet number, *Pe*, is a straightforward dimensionless expression:

$$Pe = \frac{vl}{D},$$ (1.1)

where v is flow speed, l is transport distance, and D is the diffusion coefficient. (Confusingly, a heat-transfer version of the Péclet number may be more common than this mass-transport form; it puts thermal diffusivity rather than the molecular diffusion coefficient in its denominator.)

Calculating values of the Péclet number can give us more than merely a way to check up on the performance of the evolutionary process. Often it can test hypotheses about the primary function of various features of organisms—"primary" in the sense of being most constraining on design. Perhaps that justification can be best put as a series of examples, which will follow after a few words about the origin of the ratio.

One can view the Péclet number several ways. The simplest sees it as the ratio of a convective or flow rate, v, to a diffusion rate, D/l. A slightly more formal version combines a simple numerator, mv, for flowing momentum, with a denominator that represents a simplified form of Fick's first law for diffusive momentum transport, DSm/V, where S is cross-sectional area and V is volume. Taking l^2 as a crude proxy for area and l^3 for volume, one gets equation (1.1).

From a slightly different viewpoint, the Péclet number represents the product of the Reynolds number (Re) and the Schmidt number (Sc). The first,

$$Re = \frac{\rho l v}{\mu},$$ (1.2)

where ρ and μ are fluid density and viscosity respectively, gives the ratio of inertial to viscous forces in a flow. At high values, bits of fluid retain a lot of individuality, milling turbulently as in a disorderly crowd; at low values, bits of fluid have common aspirations and tend to march in lock step formation. In short, Reynolds number characterizes the flow. The Schmidt number,

$$Sc = \frac{\mu}{\rho D},$$ (1.3)

is the ratio of the fluid's kinematic viscosity (viscosity over density, a kind of relative viscosity) to the diffusion coefficient of the material diffusing through it. It gives the relative magnitudes of the mobilities of bulk momentum (solution flow) and molecular mass (solute diffusion). In short, it characterizes the material combination, solute with solvent, that does the flowing and diffusing.

Of course the way we've swept aside all geometrical details puts severe limits on what we can reasonably expect of numerical values of Pe and should be kept in mind. Only for comparisons among geometrically similar systems can we have real confidence in specific numbers. Furthermore, we'll ignore the tacit requirement that, strictly speaking, diffusion and flow should be in the same direction. Still, because living systems vary so widely in size, even order-of-magnitude values should be instructive.

As a quick illustration of the way a value of the Péclet number can shed light on a problem, consider the way we all-too-often demonstrate diffusion for students. One drops a crystal of some soluble material whose solution is intensely colored into a container of water, and the class watches the spread of colorant through the container. Some years ago I wrote a short diatribe about the scheme in response to a published recommendation for its use (Vogel 1994a). I claimed that such demonstrations were fraudulent, that they relied on convection rather than diffusion for material transport, and in part based my case on Péclet number. Here's the argument, put more pointedly in the present context:

Imagine that one can notice diffusion in a liquid when it has transported something about a millimeter. Assume a diffusion coefficient of

10^{-10} meters squared per second, corresponding to a non-electrolyte with a molecular weight of about 100. In that case a flow speed of a mere 1 micrometer per second yields a Péclet number of 1.0 (eq. 1.1). So at any higher speed, convection will be a greater transport agent than diffusion. 1 micrometer per second—8.6 centimeters per day—that's a glacial rate in an unusually literal sense. Under no easily contrived circumstances can water in a fully liquid state be kept still enough to meet such a criterion!

Now for a few more biological cases where calculating a Péclet number might prove instructive.

INSIDE JOBS

The Sizes of Our Capillaries and Kidney Tubules

Consider our circulatory systems, in particular the size of the ultimate vessels, capillaries, where function depends both on diffusion and on flow. Do we make capillaries of the proper size—or, to be less judgmental, can we rationalize their remarkably invariant size? Efficient operation ought to be important. After all, we devote about 6.5 percent of our body volume to blood and expend about 11 percent of our resting metabolic power pushing it around. Apparently we do size them properly. A capillary radius of 3 micrometers, a flow of 0.7 millimeters per second, and a diffusion coefficient (assuming oxygen matters most) of 18×10^{-10} meters squared per second yield a Péclet number of 1.2. If anything, the value turns out a bit closer to 1.0 than one expects, considering its underlying approximations (Middleman 1972).

Of course nature might pick different combinations of radius and flow speed without offending Péclet. (We'll ignore the issue of the fit of red blood cells to capillaries by tacitly assuming an evolutionarily negotiable RBC size.) Smaller vessels would permit faster flow and lower blood volume, but the combination would (following the Hagen-Poiseuille equation) disproportionately increase pumping cost. Larger vessels would entail greater blood volume, the latter already fairly high, and slower flow, which would make the system less responsive to changes in demand. One suspects something other than coincidence behind the similarity of our relative blood volume, 6.5 percent, to that of an octopus, 5.8 percent (Martin et al. 1958).

Quite likely this choice of capillary size, based on Péclet number and some compromise of volume versus cost, sets the sizes of the other vessels of our circulatory systems in a cascade of consequences. According to Murray's law (LaBarbera 1990), the costs of construction and operation set the relative diameters of all vessels. Thus, a factor that sets vessel diameter at one level in their hierarchy ends up determining the diameters of all the rest.

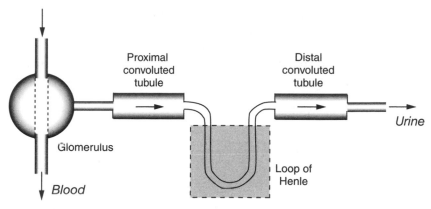

Figure 1.1. An unusually diagrammatic view of the filtration and reabsorption stages of most vertebrate kidneys.

The rule is a simple one—branching conserves the cubes of the radii of vessels, so the cube of the radius of a given vessel equals the sum of the cubes of the vessels at some finer level of branching that it supplies or drains.

What about the reabsorptive tubules of our kidneys, in particular those just downstream from the glomerular ultrafiltration apparatus (the "proximal convoluted tubules"), as in figure 1.1. We're again looking at a system that represents a far-from-insignificant aspect of our personal economies. Twenty to 25 percent of the output of the heart passes through this one pair of organs. About 20 percent of the plasma volume squeezes out of the blood in a pass, in absolute terms around 60 milliliters per minute per kidney. Each kidney consists of about 2 million individual units, the nephrons, operating in parallel. Thus each glomerulus sends on for selective reabsorption about 0.5×10^{-12} cubic meters per second.

The sites of the initial phase of reabsorption are these proximal tubules, each about 40 micrometers in inside diameter. Combined with the earlier figure for volume flow, that determines a flow speed of 0.40 millimeters per second. So we have speed and size. Diffusion coefficient can be assigned no single number, since the tubules reabsorb molecules from small organic molecules and ions to small proteins with molecular weights of around 40,000. Coefficients most likely range from about 0.75×10^{-10} to 40×10^{-10} meters squared per second. That produces Péclet numbers from 2 to 100.

At first glance these seem a bit high, but the story has another part. The calculation uses the velocity of the filtrate as it enters the tubules from the glomeruli. The tubules, though, reabsorb at least 80 percent of the volume of the filtrate, so by the time fluid leaves them, its speed has dropped by at least a factor of 5. That gives exit Péclet numbers a range of 0.4 to 20, with an average in between—quite reasonable values, indicative (to be presumptuous) of good design. The cost of pushing flow through the tubules is low, at least relative to the power requirements of filtration and of the kidney's chemical activities. So one might speculate that the system is contrived to bias its Péclet numbers so they exceed one, albeit not by much, for most molecules over most of the lengths of the tubules.

The Sizes of Plant Cells and Algal Colonies

One can argue that the boundary between the cellular and the super- (or multi-) cellular world reflects the upper size limit of practical, diffusion-based systems. In short, for anything but exceedingly leisurely large-scale life to get above the typical cell size requires some kind of convective augmentation for moving material. That makes the cellular world a diffusion-based one and the supercellular world a convection-augmented one. I like the view because it tickles my particular bias toward physical determinants.

But I have to admit that it won't work for plant cells. On average, the cells of vascular plants run about ten times the size of animal cells, taking "size" as typical length. They are of the order of 100 micrometers in length if somewhat less in width, so 25 micrometers should be typical of the distance from cell wall to center. That increased size might have devastating consequences for transport were it not for the internal convection common to such cells. Put another way, in plants the size scale at which convective transport comes into play doesn't correspond to the sizes of plant cells.

That bulk flow system within plant cells goes by the name "cyclosis." We know quite a lot—but far from all—about how microfilaments of actin (a key component of muscle) power it, but here only its speed matters. That speed is around 5 micrometers per second (Vallee 1998). Focusing on oxygen penetration and using a penetration distance of 25 micrometers gives a Péclet number of 0.07. That tells us that the system remains diffusion dominated, that cyclosis doesn't reach a significant speed to make much difference to transport effectiveness. Looking at carbon dioxide penetration, with a diffusion coefficient of 0.14 meters squared per second, raises that number too little to change the conclusion. So why bother with cyclosis?

Perhaps we should take a different view. From size, speed, and a presumptive Péclet number around 1 we can calculate a diffusion coefficient of 1.25×10^{-10} meters squared per second. That corresponds to a non-ionized molecule with a molecular weight of about 6000. Thus plant cells appear diffusion dominated for dissolved gases, amino acids, sugars, and the like. But they make significant use of (in effect, need) convection for moving proteins and other macromolecules, where molecular weights are in the thousands. In a sense, of course, that reliance on cyclosis bolsters the underlying argument about the practical size limit of diffusional systems.

Essentially the same picture emerges from a different (and much more sophisticated) analysis, one by Pickard (2006). The situation of an organelle provides the vantage point here. If stationary in the cytoplasm, diffusion limits its effective rates of absorption or secretion, but cyclosis even at 1 micrometer per second would produce sufficient convective augmentation to double transport rates. That point, $Pe = 1.0$, corresponds to an organelle size of 1 micrometer and a diffusion coefficient of 10^{-12} meters squared per second—a molecular weight of about 10,000,000. It establishes another transition point as well. For values below 1.0, flux increases with Pe^2, while for higher values, flux increases only with $Pe^{1/3}$. Pickard (2006) also provides an especially good entry into the nonbiological literature on Péclet number.

That boundary between cellular and multicellular worlds also receives scrutiny in the work on *Volvox* by Solari et al. (2006) that was mentioned earlier. Mature colonies apparently require flow for adequate growth of gonidia and the daughter colonies within their lumens. One might imagine that inverting each cell so flagella faced inward rather than outward would provide proper internal circulation—the equivalent of cyclosis. But the flagella also act as propulsive devices that move the colonies around, preventing sinking and enabling them to explore spatial variation in nutrient concentration. Still, putting them on the outside does help the situation. Augmenting flow around the outside assures minimal local depletion and gives internal diffusive transport a better starting point. The report gives Péclet numbers in the hundreds, but these take colony diameter as characteristic length. I think a more appropriate length for present purposes is that of the flagella, about 12 micrometers, the length over which convection and diffusion might well balance. Combining that with a swimming speed of 400 micrometers per second and a diffusion coefficient of 2×10^{-10} meters squared per second gives $Pe=24$. So the system looks mildly convection-biased, which may offset the lack of internal flow and the resulting reliance of daughter colonies on diffusion within their parents.

TRANSPORT AT EXTERNAL SURFACES

Sinking Speeds of Phytoplankton

Diatoms plus some other kinds of small algae account for nearly all the photosynthetic activity of open oceans. Paradoxically, most of these light-dependent phytoplankters are negatively buoyant most of the time.

Not that they sink rapidly; 4 micrometers per second (a foot a day, in antediluvian everyday units) is typical. According to one common explanation, such sinking improves access to carbon dioxide by minimizing the organism's own local depletion of dissolved gas. In effect, the cell walks away from its personal environmental degradation. Still better, it walks away without locomotory cost. Of course it (or its progeny) will eventually suffer, inasmuch as sinking takes it to depths at which respiratory demand exceeds photosynthetic rate. Somehow (and wave-induced water mixing comes into the picture) some organism-level cost-benefit analysis favors this slight negative buoyancy.

Calculating a Péclet number casts serious doubt on the notion of escape from local CO_2 depletion, doubt long ago raised (with an equivalent argument) by Munk and Riley (1952). From that sinking rate of 4 micrometers per second, the diffusion coefficient of CO_2, 14×10^{-10} meters squared per second, and as distance the 10-micrometer diameter of a typical diatom, we get a value of 0.03. So diffusion rules; convection, here due to sinking, will not significantly improve access to carbon dioxide. We might have chosen a larger distance over which CO_2 had to be transported to be available at an adequate concentration, but even a distance ten times greater would not raise Pe enough to pose a serious challenge to the conclusion.

Why, then, should a phytoplankter sink at all? The argument tacitly assumed uniform concentration of dissolved gas except where affected by the organism's activity. It left open the possibility that a diatom might be seeking regions of greater concentration, even lowering its sinking rate wherever life went better, something mentioned earlier when considering the much larger *Volvox*. In a world mixed by the action of waves, that's uncertain, although patchiness isn't unknown and (as appears the case) buoyancy does in fact vary with the physiological state of a cell. Perhaps phytoplankters bias their buoyancy toward sinking so they won't rise in the water column and get trapped by surface tension at the surface. If perfect neutrality can't be assured, then sinking may be preferable, provided the speed of sinking can be kept quite low—as it is. Surface tension may be a minor matter for us, but it looms large for the small. In the millimeter to centimeter range a creature can walk on it since the Bond number, the ratio of gravitational force to surface tension force, is low. A smaller creature may not be able to get loose once gripped by it; specifically, the Weber number, the ratio of inertial force to surface tension force may drop too far (Vogel 1994b). But risky surface entrapment presumes that the surfaces of these diatoms and other small organisms are fairly hydrophobic, which, I'm told, may not be the case. So another hypothesis would be handy—plus some experimental work on interactions of individuals with interfaces. Calculating Péclet, Bond, and Weber numbers certainly raises a possibility that at least ought to be ruled out.

I should note that a somewhat different version of the Péclet number has come into use in studies of natural populations of phytoplankton. Instead of molecular diffusivity, it uses so-called eddy diffusivity, an effective diffusivity set by a combination of turbulent mixing (mainly) and molecular diffusivity (additionally). Values of D run around 10^{-5} to 10^{-3} (MacIntyre 1993) rather than the 2×10^{-9} and down for molecular diffusion in water. A useful paper with copious references is O'Brien et al. (2003).

Swimming by Microorganisms and Growing by Roots

We most often think of movement by active swimming rather than by passive sinking. Some years ago, Edward Purcell (1977), a physicist, wrote a stimulating essay about the physical world of the small and the slow, looking in particular at bacteria. Among other things, he asked whether swimming, by, say, *Escherichia coli*, would improve access to nutrients. (Yes, that's the full name of *E. coli*.) By his calculation, a bacterium 1 micrometer long, by swimming at 20 micrometers per second (see Berg 1993) would only negligibly increase its food supply, assuming it to be dissolved sugar. To augment its supply by a mere 10 percent, it would have to go no less than 700 micrometers per second. That's well above the fastest swimming speed I've encountered for a bacterium—about 140 micrometers per second for a free-living marine species, *Vibrio harveyi* (Mitchell et al. 1995).

Purcell's answer to the question of why swim at all turned on the heterogeneity of ordinary environments and the advantage of seeking the bacterial equivalent of greener pastures, as suggested above for diatoms. That rationalization, incidentally, receives support from a recent literature on a sort of micro-patchiness within macroscopically uniform liquid environments. Otherwise the bacterium resembles a cow that eats the surrounding grass and then finds it most efficient, not to walk, but to stand and wait for the grass to grow again.

The Péclet number permits us to cast the issue in more general terms. Sucrose has a diffusion coefficient of 5.2×10^{-10} meters squared per second; together with the data above we get a Péclet number of about 0.04. Swimming, as Purcell said, should make no significant difference. But the conclusion should not be general for microorganisms. Consider a ciliated protozoan, say *Tetrahymena*, which is 40 micrometers long and can swim at 450 micrometers per second. If oxygen access is at issue, the Péclet number comes to 10, indicating that swimming helps a lot. Indeed it might just be going unnecessarily fast, prompting the thought that it might swim for yet another reason—or reasons.

Growing roots provide a case just as counterintuitive as that of swimming bacteria but in the opposite direction (Kim et al. 1999). A root can affect nutrient uptake by altering local soil pH. Root elongation speeds run (perhaps an inappropriate word) around 0.5 micrometers per second, again down in the range of a glacier not yet goaded by global warming. But it turns out to constitute a significant velocity, enough so that (at least in sandy soil) the Péclet number gets well above 1. Taking root diameter as length, Pe values for rapidly diffusing H^+ ions may exceed 30. Thus motion most likely affects the pH distribution more than does diffusion in the so-called rhizosphere.

Flow Over Sessile Organisms

For sinking diatoms and swimming microorganisms we asked about why creatures did what they did. In some other situations we can test claims about the physical situations in which they live, in particular about local flows. How fast must air or water flow over an organism to affect exchange processes significantly? To put the matter in sharper terms, can calculating Péclet numbers help us evaluate a claim that extremely slow flow matters? After all, neither producing nor measuring very low speed flows is the most commonplace of experimental procedures so, at the least, cited speeds should be viewed with a skeptical eye.

For instance, consider the claim that a flow of 10 millimeters per second significantly increases photosynthesis in a green alga, *Spirogyra* (plate 1.2 left), consisting of threadlike filaments about 50 micrometers in diameter (Schumacher and Whitford 1965). Inserting the diffusion coefficient of CO_2 gives a Péclet number around 100 and suggests that far slower flows should also matter. Now one wonders about the opposite issue, whether so-called still water, the control in such comparisons, is still enough to achieve truly negligible flow. My own experience suggests that thermal convection and persistence of currents left from the initial filling of a tank can complicate attempts to prevent water from flowing. Still water doesn't just happen, and that may afflict laboratory investigations as well as classroom demonstrations.

Another paper reports that a flow of 0.2 to 0.3 millimeters per second, about a meter per hour, significantly increases photosynthesis in an aquatic dicot, *Ranunculus pseudofluitans* (Westlake 1977; plate 1.2 right). Its finely dissected, almost filamentous, leaves are about 0.5 millimeters across. A Péclet number of about 300 gives credibility to an otherwise eyebrow-raising report. One again guesses that even slower flows should be significant. The tables turn—one now becomes skeptical of any casual assumption of effectively still water in ponds and lakes.

A third paper (Booth and Feder 1991) considers the influence of water flow on the partial pressure of oxygen adjacent to the skin of a salamander, *Desmognathus*. The authors found that currents as low as 5 millimeters per second increased that partial pressure, facilitating cutaneous respiration. With a diameter of 20 millimeters, that flow produces a Péclet number of 50,000. Even assuming instead a 1-millimeter thick mixed regime at its skin gives a Péclet number of 5,000. A sessile *Desmognathus* may need flow, but it surely doesn't need much. Once again, the quality of any still-water control becomes important—something best checked, perhaps by watching a blob of colorant, before submitting a paper explicitly or tacitly assuming it.

Two Functions for Gills

Many aquatic animals both respire, exchanging dissolved gases, and suspension feed, extracting edible particles from the surrounding waters. Structures such as gills, with lots of surface relative to their volumes, can perform either function. While most suspension-feeding appendages may look nothing like gills, some not only look like gills but share both name and functions. No easy argument suggests that bifunctional gills balance their two functions. Quick calculations of Péclet number can tell us which function dominates a particular design and can thus point to the features that distinguish a respiratory gill from a dual-function gill.

Consider a keyhole limpet, *Diodora aspera*, a gastropod mollusk that uses its gills for respiration. With gill filaments about 10 micrometers apart, a flow rate of 0.3 millimeters per second (according to Janice Voltzow), and the diffusion coefficient for oxygen, the Péclet number comes to about 2. That's about as good as it gets for a respiratory gill. The dual-function gills of a bivalve mollusk, the mussel *Mytilus edulis*, contrast sharply. They have an effective distance of about 200 micrometers and a flow speed of about 2 millimeters per second (Nielsen et al. 1993). For oxygen access, that gives a Péclet number around 100. Clearly, mussel gills pump far more water than would be necessary were respiration the design-limiting function.

One can do analogous calculations for fishes, a few of which use gills for suspension feeding as well as respiration. A typical teleost fish has sieving units 20 micrometers apart (Stevens and Lightfoot 1986) with a flow between their lamellae of about 1 millimeter per second (calculated from the data of Hughes 1966). For oxygen transport, the resulting Péclet number is 5.5, not an unreasonable value for an oxygenating organ. One gets quite a different result for a fish that uses its gills for suspension feeding. While a somewhat higher 80 micrometers separates adjacent filtering elements, the main difference is in flow speeds. Flows run around 150 millimeters per second for passive ("ram") ventilators (Cheer et al. 2001) and 550 millimeters per second for pumped ventilators (Sanderson et al. 1991). The resulting Péclet numbers, 6,500 and 20,000 (again using oxygen diffusion), exceed anything reasonable for respiratory organs.

Air Movement and Stomatal Exchange

All the previous cases looked at diffusion and convection in liquids. Of course the same reasoning should apply to gaseous systems as well—fluids are fluids, and diffusion and convection happen in both phases of matter.

Leaves lose or "transpire" water as vapor diffuses out though their stomata and disperses into the external air. Transpiration rates depend on a host of variables, among them wind speed and stomatal aperture, with this last under physiological control. Right next to a leaf's surface, the process depends, as does any diffusive process, on concentration gradient—here from the saturated air at the stomata to whatever the environmental humidity might be. The stronger the wind, the steeper the concentration gradient as the so-called boundary layer gets thinner.

Consider a bit of leaf 20 millimeters downstream from the leaf's upwind edge. Assume a wind about as low as air appears to move for any appreciable length of time, as my slightly educated guess, 0.1 meter per second. The effective thickness of the velocity gradient outward from the leaf's surface, δ, can be calculated using the semi-empirical formula (Vogel 1994b)

$$= 3.5 \sqrt{\frac{x\mu}{v\rho}}, \tag{1.4}$$

where x is the distance downstream, and μ and ρ are the air's viscosity and density, respectively, 18×10^{-6} pascal-seconds and 1.2 kilograms per cubic meter. The thickness comes to 6 millimeters. (The datum must be regarded as the crudest approximation; among other things, the formula assumes a thickness that is much less than the distance downstream.) For that thickness, that wind speed, and the diffusion coefficient of water vapor in air, 0.24×10^{-4} meters squared per second, the Péclet number is 25. So even at that low speed, wind suffices to produce a convection-dominated system.

What might that tell us? For one thing, it implies that changes in wind speed should have little or no direct effect on water loss by transpiration. If water loss does vary with wind speed, one should look for something other than a direct physical effect, something such as changes in stomatal aperture. For another thing, it implies that a leaf in nature won't have adjacent to its surface very much of a layer of higher-than-ambient humidity.

So-called vapor caps are not likely to mean much with even the most minimal of environmental winds. It also rationalizes observations that changes in stomatal aperture area have considerable effects on transpiration rates—vapor diffuses through them, and in an otherwise convection-dominated process, diffusion becomes rate-limiting.

GROWTH AND DEVELOPMENT

Hydrodynamics and Growth

In a multicellular organism, developmental patterns depend on both genetic control and an organism's environmental situation, with a diversity of feedback mechanisms integrating the two. Where the environment appears to rule, one may have trouble separating direct physical effects from those mediated by active sensing and responses, linked by such feedback. But sometimes the value of a Péclet number can argue the case for direct action.

A particularly nice illustration comes from work on the accretive growth of stony corals. Like such other sessile animals as sponges, corals are suspension feeders, gathering food over their external surfaces. The effectiveness of feeding depends on overall form, whether of individuals, as in sponges, or of colonies, as in corals. And intraspecific form depends strongly on the local situation—particularly on currents around individuals or colonies. In the stony corals, colonies of branching forms from quiescent locations tend to have more open branches with thinner and, eventually, longer branches. By contrast, colonies of a given species subjected to higher average flows are considerably more compact and spherical.

One might expect the degree of compactness to track the Reynolds number of flow in the immediate habitat. But a series of simulations (Kaandorp and Sloot 2001; Merks et al. 2003) found that variation in Péclet number much better describes what happens. So flow—advection—relative to diffusion, not just flow per se, must be important. One might then wonder about the meaning of diffusion where the key items are motile microorganisms such as unicellular algae. These investigators found that they could use effective diffusion coefficients for such ingestibles, with values, figured from several starting points, well above those of even the smallest of molecules. For instance, the green alga *Chlamydomonas*, 10 micrometers in diameter, has an effective D of 5×10^{-8} meters squared per second, and a 1-millimeter mover, a D of 3.5×10^{-5}. So it looks as if an analysis based on Péclet number has value even under circumstances in which the variable might seem inapplicable. And a particular abiotic mechanism provides a sufficient (if not necessarily complete) explanation of quite a lot of morphological variation.

Another somewhat counterintuitive way the mix of diffusion and convection affects growth has been uncovered in nonmotile phytoplankton. These, of course, move passively with the surrounding water, eliminating an obvious role for local flow. Moreover, the flows in which they're embedded will be laminar rather than turbulent, simply as a matter of scale. The minimum size of a turbulent eddy depends on the rate of energy dissipation in a fluid and is described by the so-called Kolmogorov scale. For natural waters it runs around 1 millimeter, roughly the upper end of the size range of phytoplankton. Thus, they should see little if any of the cross-flow mixing concomitant with turbulence. But the strong local velocity gradients that they will experience will make them rotate and even cause such detrimental things as cell aggregation and cell destruction (Hondzo and Lyn 1999). The same local, small-scale flows can still facilitate growth by enhancing nutrient uptake—mainly when Péclet numbers exceed unity (Warnaars and Hondzo 2006).

Chemosensory Systems and Yet Other Possible Relevancies

Where else might calculations of Péclet numbers provide useful insight? We haven't considered, for instance, olfactory systems, either aerial or aquatic—in short, chemosensation in moving fluids. For these, I've found few relevant data. Several things have come up recently that suggest anything but the kind of match between convection and diffusion that we've been treating as in some way or another optimal. But that mismatch may be instructive.

A look at Mead (2005) or Woodson et al. (2007) will provide a good sense of what goes on, at least in relevant aquatic systems. Fishes most often detect dissolved odorants in a pair of chambers atop their heads, just behind their—to use the term figuratively—noses. By some mix of active pumping and hydrodynamic induction, in each water flows in an anterior opening and out a posterior one. Cox (2008) calculated Péclet numbers between 20 and 3,000, indicating severe biases toward convection. Decapod crustaceans do chemoreception with quite different machinery, locating the receptors ("aesthetascs") on their antennules, flicking these to improve effective contact with odor-laden water. (See, for instance, Mead and Weatherby 2002.) For crayfish, Kristina Mead (personal communication) calculates Pé's in the thousands.

Perhaps two factors, in combination, underlie that bias. First, as Cox (2008) points out, the cost of pumping water should be insignificant relative to overall metabolic rates. Furthermore, pumping ought to reduce the lag time between encountering signal-carrying water and detecting the chemical information. And whether looking for prey or looking out for predators, response speed should be accorded very high priority.

Beyond chemosensation, one wonders about yet other systems. Might we usefully consider the speeds and distances of movement of auxins and other plant hormones? Might we learn anything by comparing systems in which oxygen diffuses within a moving gas with ones in which it diffuses in a flowing liquid? Here one thinks of systems such as, on the one hand, our alveoli and bronchioles (see, for instance, Federspiel and Fredberg 1988), the tubular lungs of birds, and the pumped tracheal pipes of insects and, on the other, the gills of fish, crustaceans, and the like.

The Sizes of Morphogenetic Fields and Synaptic Clefts

A variation of the Péclet number may provide insight into such things as the development of animals and characteristic biological times. Much of pattern formation depends on the diffusion of substances, morphogens, whose concentration gradients establish embryonic fields. Establishing larger fields not only means lower gradients (or takes higher concentrations of morphogens) but (for a given concentration) would take more time, a non-negligible resource in a competitive world. Breaking up velocity into length over time we get:

$$\frac{l^2}{Dt}. \tag{1.5}$$

(The reciprocal of this expression is sometimes called the mass transfer Fourier number.)

To define the limits of excessively diffusion-dependent systems, we might assume a value of 1. A typical morphogen has a molecular weight of 1000; its diffusion coefficient when moving through cells (a little lower than through water) ought to be around 1×10^{-10} meters squared per second. A reasonable time for embryonic processes should be a few hours, say 10^4 seconds. The numbers and the equation imply that embryonic fields should be around 1 millimeter across, about what we find. I think this basis for the embryonic field size (although put somewhat differently) was first argued by Crick (1970).

In effect, the calculation produces what we might consider a characteristic time for a diffusive process. Consider ordinary synaptic transmission in a nervous system. The most common transmitter substance, acetylcholine, has a molecular weight of 146 and a diffusion coefficient around 7×10^{-10} meters squared per second. With a synaptic cleft of 20 nanometers, the corresponding time comes to 0.6 milliseconds. That value matches cited values for overall synaptic delay, between about 0.5 and 2.0 milliseconds, consistent with the textbook attribution of much of the delay to transmitter diffusion.

That Diffusion Coefficient

So far, values of the diffusion coefficient have materialized ex cathedra. In practice, one obtains them just that way—by appeal to higher authority or whoever bequeaths us the appropriate tables. That larger molecules have lower coefficients needs no further emphasis; it does tacitly anticipate some proper formula. No such luck, at least for ordinarily reliable predictions, but quantitative guidance isn't completely lacking. Uncertainty comes from the dependence of diffusion coefficients (and mean free paths of molecules, on which they directly depend) on molecular shape, no easy thing to specify. The following formula gives a rough sense of how the coefficient varies with molecular (here as molar) mass, assuming that the diffusing molecules are spherical, that the solution is quite dilute, and that some other conditions hold:

$$D = \frac{KT}{6\pi\mu r} = \frac{KT}{6\pi\mu} \sqrt[3]{\frac{4\pi\rho N}{3m}}. \qquad (1.6)$$

Here K is the Boltzmann constant, T the Kelvin temperature, μ the viscosity of the solvent, r the molecular radius, ρ the density of the solute material, N Avogadro's number, and m the molecular weight of the solute.

Put as a rule of thumb, the diffusion coefficient varies inversely with the cube root of the molecular weight of the diffusing substance. In practice we get our diffusion coefficients from empirical measurement and thence, by inference, some idea of molecular shape. Incidentally, diffusion coefficients for transport through living material such as muscle and connective tissue are anomalously low (Schmidt-Nielsen 1997).

In fields such as fluid mechanics and chemical engineering, dimensionless numbers are pervasive and of proven utility. Biologists have been slow to exploit them as tools, perhaps because our initial or sole contacts with physical science have been with physicists, who use them only rarely. Or perhaps because we've for so long focused on scaling relationships, with their irregular and ambiguous dimensions. Dimensionless numbers will turn up again and again in the chapters that follow, in part as my argument-by-example (as briefly in Vogel 1998) that they can help us see the relevance of physical phenomena to biological systems. Some, such as the present Péclet number, have been bequeathed to us by engineers (here mainly chemical engineers), and we needn't even redefine the variables. Others (the Froude number, for instance, which will appear in chapters 6 and 7) have traditional applications in engineering that we can, by such redefinition, put to quite different uses. We can generate still other dimensionless ratios without great formality, bearing in mind that the ultimate—indeed, the only relevant—test is whether the number gives some useful insight, perhaps permitting us to discern general relationships that would otherwise remain obscure.

Dimensionless numbers provide tools that, for the most part, tell us what matters. Is a flow going to be laminar or turbulent? At a given speed will an animal of a certain size find it easier to walk or run? Or, as here, will convective or diffusive processes prove predominant, so one or the other can be ignored, or does one's system appear designed to take best advantage of each? These chapters will argue (again by example) that these ratios prove especially useful where systems encompass wide size ranges. In particular, by not having any apparent or concealed length dimension, they can avoid contamination of comparative numbers with the confusing effects of size per se.

Finally, who was this person Péclet? One does not normally name a number after oneself. Someone may propose a dimensionless index, and then the next person who uses it names it after the first. Or else its originator may name it for some notable scientist who worked in the same general area. Péclet number is a case of the latter. Jean Claude Eugène Péclet (1793–1857) was part of the flowering of French science just after the revolution. He was a student of the physical chemists (as we would now call them) Gay-Lussac and Dulong—names yet remembered for their laws—and a teacher of physical science. He did noteworthy experimental work on thermal problems and wrote an influential book, *Treatise on Heat and its Applications to Crafts and Industries* (Paris 1829).

Putting his name on a dimensionless number happened a century later, by Heinrich Gröber, in 1921, in another important book, *Fundamental Laws of Heat Conduction and Heat Transfer*. That thermal version of the Péclet number antedates the mass-transfer version used here. The latter, as far as I can determine, first appears in a paper on flow and diffusion through packed solid particles, by Bernard and Wilhelm, in 1950. They note its similarity to the dimensionless number used in heat-transfer work and call their version a "modified Peclet group, symbolized Pe'". They shift, confusingly and deplorably, from an acute accent in "Péclet" to a prime ('), now usually omitted, at the end. Analogous indices for thermal and material processes are not unusual since the underlying transport processes are either the same (as convection) or analogous (as diffusion and conduction). But most often the two carry different names—such as Prandtl number and (as earlier) Schmidt number. Amusingly, most sources mention one of the versions of the Péclet number with no acknowledgment that there is any other.

CHAPTER 2

The Bioballistics of Small Projectiles

Ours has long been a projectile-ridden culture, perhaps from our days of throwing rocks and hurling spears in efforts to damage potential edibles, predators, and competitors—the latter including conspecific ones. Long before inventing firearms, we made devices that improved the effectiveness of projectiles—atalatls, archery, catapults, and so forth. We rarely think of projectiles in any nonhuman context, but they turn out to be both common and diverse. Many animals jump; many plants shoot their seeds. While "many" may not imply "most," terrestrial life is rich with examples of ballistic motion, motion in which a projectile gets all of its impetus prior to launch.

For most of us, projectiles and their trajectories played only a brief and unmemorable role early in a basic physics course. Some conveniently tidy equations emerged (often as homework) in unambiguous fashion from just two facts. The horizontal speed of a projectile remains constant; only the downward acceleration of gravity (g) alters its initial vertical speed. Where launch and landing heights are the same, a simple formula links range (d) with launch speed (v_o) and projection angle (θ_0) above horizontal:

$$d = \frac{v_o^2 \sin 2\theta_o}{g}. \tag{2.1}$$

Thus, for a given initial speed, a projectile goes farthest when launched at an upward angle of 45°. That maximal range is simply

$$d_{max} = \frac{v_o^2}{g}. \tag{2.2}$$

So an initial speed of 40 meters per second (144 kilometers per hour) could take a projectile 163 meters. En route, the projectile reaches a maximum height, h_{max}, of a quarter of that best range, or

$$h_{max} = \frac{v_o^2}{4g}. \tag{2.3}$$

The trajectory forms a nicely symmetrical parabola, and the loss of range at angles above 45° exactly mirrors the loss at lower angles—as shown in

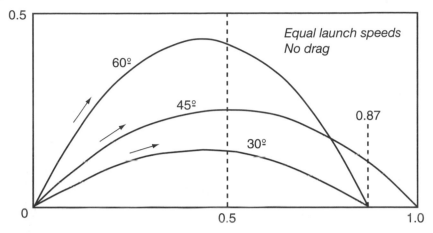

Figure 2.1. Without drag, trajectories are perfectly parabolic, with descent speeds and angles equal to ascent speeds and angles. For a given initial speed, maximum range occurs with a launch angle of 45°, and ranges after either 30° or 60° launches are 87 percent of that maximum.

figure 2.1. Such tidiness gives us biologists (as Joel Cohen, a biologist, once said) severe physics-envy.

In promoting these expressions, text or teacher may mutter, parenthetically or *sotto voce*, something about an assumed absence of air resistance, about the presumption that drag exerts a negligible effect. Nevertheless the scheme generates significant errors even for cannon balls. It gives still worse errors for golf balls—drag can halve the range of a well-driven golf ball (Brancazio 1984). The errors are tolerable only because golfers, however fanatic, rarely turn for help to physics.

What keeps a projectile going is inertia; whether we view its consequences in terms of momentum or kinetic energy, mass provides the key element. Ignoring, to take a broad-brush view, variation in both density and shape, mass follows volume. What slows a projectile are two factors, gravity and drag. The standard equations deal with the downward force of gravity and produce their nice parabolas. Drag, the force that acts opposite the direction of motion, reveals itself in deviations from such simple trajectories. Its magnitude varies in proportion either to surface area or diameter, depending on the circumstances (more specifically on the Reynolds number, about which more later). In any case, the smaller the projectile, the greater are both its surface area and its diameter relative to its volume. So the smaller the projectile, the less adequately that idealized, dragless trajectory should describe its motion. Similarly, since gravitational force, kinetic energy, and momentum all depend on mass, the less dense the projectile, the greater will be the relative influence of

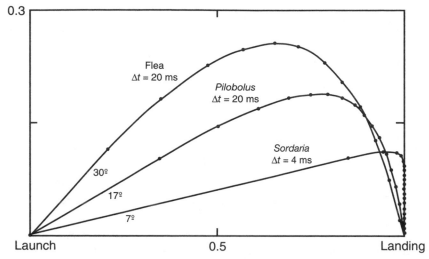

Figure 2.2. Maximum-range trajectories for three quite drag-afflicted projecties—a jumping flea, a *Pilobolus* sporangium, and a *Sordaria* eight-spore cluster. Marked points on each curve give distances after equal time increments. (Time increments differ among the three.) Note that axes give horizontal distance as a fraction of maximum range and that the *y*-axis has been expanded twofold relative to the *x*-axis.

drag. Thus the smaller and less dense the projectile, the less parabolic will be its trajectories, and the more divergent will be its launch and landing speeds, as one can see in figure 2.2.

The upshot is that these simple equations usually perform poorly for biological projectiles. Few are very large and none very dense, so their performances pale beside those of long-traveling and damage-inducing chunks of rock or iron. Nonetheless, their diversity in ancestry, size, and function makes life's projectiles worth some attention. Sports, hunting, and warfare, the uses that come first to mind, matter least often to species other than our own. Instead, two functions dominate. Some organisms jump, forming single, whole-body projectiles; others shoot propagules—fruits, seeds, spore clusters, even individual spores.

DEALING WITH DRAG

To look with any degree of realism at the trajectories of biological projectiles, we must, so to speak, put drag into the equation. As it happens, that turns out to be trickier than one might expect. We biologists often

imagine a physical world run according to straightforward (if sophisticated) rules, at least when compared with the messy scene that evolution
generates. At least for a simple object such as a sphere, drag ought to behave with predictable lawfulness rather than with our eccentric awfulness.
One should be able to look up a basic equation for drag as a function of
speed or size. Not so! Over the range of speeds and sizes that might matter
to organisms, these are distinctly ill-tempered functions. The trouble comes
from changes, sometimes abrupt, in how fluids flow over objects—whether
laminar or turbulent, whether surface-following or separated, and so forth.
For a large object going at a fairly high speed, drag varies with the square
of speed and with the area of the object. For a small object going slowly,
drag varies directly with speed and with the length of the object. In between, the relationship bears no resemblance to anything that tempts application of our customary regressions and power laws fittings.

Fortunately, two twentieth-century accomplishments save the day. First,
from direct measurements we know how drag varies with speed and size
for ordinary objects such as spheres moving through ordinary fluids such
as air and water. And, second, even the most minimal desktop computer
now makes short work of calculating draggy trajectories by an iterative
approach. One starts with a projectile of a given size, density, launch speed
("muzzle velocity" in the parlance of these violent times), and launch angle. After a short interval, the computer informs us of the projectile's
slightly different speed and path, the two altered by gravity, acting downward, and drag, acting opposite the projectile's direction of motion. The
computer then takes the new speed and path as inputs and repeats the calculation to get yet another speed and path. In the simplest case, the computer stops iterating when the projectile's height has returned to that of its
launch—when it has returned to the ground.

The way drag gets into the picture, though, takes a little explanation.
We normally express drag in dimensionless form, as the so-called drag coefficient, C_d. It amounts to drag (D) relative to area (S) divided by a
kind of idealized pressure, that which would push on something were the
fluid coming directly at it to effect a perfect transfer of momentum and
then obligingly (and quite unrealistically) disappear from the scene.
Specifically,

$$C_d = \frac{D/S}{\rho v^2/2} \tag{2.4}$$

where ρ is fluid density and v is the speed of the object through the fluid.
The commonest reference area is the maximum cross-section of the object normal to flow, the area facing the oncoming fluid. Unfortunately, the
relationship between speed and drag coefficient behaves little better than
that between speed and drag itself—the equation just dedimensionalizes

drag. If drag were simply proportional to area, fluid density, and the square of velocity, then C_d would be constant (and unnecessary). So variation in C_d exposes the eccentricities of drag. And C_d depends, not only on shape, but on the object's size and on the fluid's viscosity (μ here, but often η) and density.

Fortunately, these last three variables operate as a particular combination, the dimensionless Reynolds number, mentioned earlier and in the previous chapter,

$$Re = \frac{\rho l v}{\mu}, \tag{2.5}$$

where flow-wise length provides the commonest reference l. Again, Re represents the ratio of inertial to viscous forces as fluid crosses an immersed object. Non-intuitive and still untidy perhaps, but now one needs to know only how drag coefficient varies with Reynolds number, and all the other relationships follow, at least for a given shape.

For present purposes this last function, $C_d = f(Re)$, breaks into three separate domains. When Re exceeds 100,000 (assuming a sphere), C_d is very nearly 0.1. For Re's between 1,000 and 100,000, C_d is about 0.5. Thus for both domains, drag varies with the square of speed, but with different constants of proportionality. For Reynolds numbers from 1 to 1,000, the best encapsulation I've seen comes from White (1974):

$$C_d = \frac{24}{Re} + \frac{6}{1 + Re^{1/2}} + 0.4. \tag{2.6}$$

(The first term on the right side represents Stokes' law, commonly used to calculate sinking rates of tiny spherical particles or bubbles in air or water. One doesn't need the other terms at Reynolds numbers below about 1.) The computer has only to calculate the Reynolds number to decide, for each iteration, which of the approximations to apply.

Starting with a projectile's size, density, launch angle, and launch speed, such a computational program gives all the important characteristics of a realistically draggy trajectory—range, maximum height, impact angle, and impact speed. Looking at the computation point by point gives the shape of the trajectory. With only a little playing around one can work back from an observed range to a launch speed. Of course, the scheme assumes spherical projectiles, but most non-streamlined objects can be reasonably approximated as spheres of the same (or a little greater) volume. A version of such a program can be found in appendix 2 of Vogel (1988); a version created by Tiffany Chen that displays trajectories can be downloaded from http://www.sicb.org/dl/biomechanicsdetails.php3?id=63. For anyone who cares to create a version, the program is described a bit more fully—pseudocode, essentially—as an appendix to this chapter.

As an example, consider the projectile of a particular cannon—one whose barrel (on a new carriage) graces Edinburgh Castle for the delectation of tourists. In 1457, James II of Scotland took delivery of the weapon, called "Mons Meg" after Mons, Belgium, where it was produced and an anonymous Meg (or Margaret). It may have been too heavy to work, as intended, as a transportable siege weapon, but it was certainly powerful— apparently it could throw stone spheres half a meter in diameter about 3000 meters. Assuming a typical density for stone (2700 kilograms per cubic meter), the computer program yields a launch speed of 180 meters per second and a launch angle for maximum range (for that speed) of about 43°, a shade lower than the dragless 45°. No longer do the ranges at 30° and 60° match; now the 30° range wins by about 4 percent. Drag drops the best range of the projectile to 85.9 percent of the dragless, 45° calculation—we might say that it incurs a "drag tax" of 14.1 percent. The difference might have mattered if the cannon could have been accurately aimed and ranged.

Another cannon, while they're in our sights. In 1864, toward the end of the American Civil War (in the south sometimes, still, "that recent unpleasantness"), the Confederacy imported an Armstrong cannon from England, among the last cannon that fired spherical balls. It pointed seaward at Fort Fisher, near Wilmington, North Carolina, where (after an interval in the victorious north) it now stands. This weapon could reportedly fire an iron ball, 20 centimeters in diameter and weighing 150 kilograms, 8.1 kilometers (5.4 miles). The computer tells me that without drag it would have achieved 11.9 kilometers. So, mainly on account of higher projectile speeds, the drag tax increased to 32 percent over the intervening four centuries. Optimum launch angle decreased just a bit, to 42.5°.

So the simple formulas we were taught may fall short (long, really) even where we were led to believe they applied.

PLAYING GAMES WITH BALLS

As mentioned earlier, drag bothers a well-driven golf ball. If dragless, an initial speed of 60 meters per second (216 kilometers per hour) with a 45° launch would take it 365 meters. Drag reduces that to 243 meters, a tax of no less than 35.3 percent, assuming an optimal launch angle of 41.5°. That noticeably distorts the standard parabola, with a descent a bit steeper than the preceding ascent and with a landing speed a little below launch speed. Is this result general for the balls we use in our various sports? Table 2.1 gives some data, calculated with the program just described, and figure 2.3 puts the matter in graphic form.

TABLE 2.1
Input data and simulation results for the various projectiles, from least to most drag-afflicted. Landing speeds assume launch at the angles that maximize horizontal range and assume equal launch and landing elevations. Densities, beyond ones noted in the text—1864 cannon, 7870; 1457 cannon, 2700; golf ball, 1129; basketball, 81.8; tennis ball, 394; table tennis ball, 80.6; arthropods, 500; seeds, pollen, and spores, 1100 kilograms per cubic meter.

Projectile	Effective diameter (mm)	Launch Speed (ms⁻¹)	Maximum range (m)	Launch Reynolds number	Best launch angle	Range loss from drag (%)
Kangaroo rat jump	100.	4.4	1.84	29, 000	44.5	1.1
Salticid spider jump	6	0.75	0.057	300	44°	1.4
Fruit-fly larval jump	4.5	1.2	0.14	350	44°	2.0
Bristletail jump	6	1.4	0.11	550	44°	2.5
Springtail jump	2	0.45	0.045	91	44°	4.3
Caterpillar pellet shot	2.8	1.6	0.246	300	44°	5.4
Desert locust jump	10.	3.0	0.85	2,000	44°	6.1
Click beetle jump	7	2.4	0.56	1100	44°	6.5
Box moss mite jump	0.8	0.45	0.019	24	43.5°	8.9
Aphrophora (froghopper)	6.4	3.4	1.04	1500	43.5°	11.6
Basketball	240.5	20.	35.7	320,000	43.5°	12.3
Cercopsis (froghopper)	6.4	3.8	1.3	1600	43.5°	13.6
Cannonball (1457)	500.	180.	3,000	6,000,000	43°	14.1
Lepyronia (froghopper)	4.8	4.6	1.6	1500	43%	24.0
Philaenus (froghopper)	4.1	4.7	1.6	1300	42.5°	28.6

Table 2.1 (*continued*)

Projectile	Effective diameter (mm)	Launch Speed (ms⁻¹)	Maximum range (m)	Launch Reynolds number	Best launch angle	Range loss from drag (%)
Cannonball (1864)	203	342	8,047	4,600,000	42.5	32.4
Golf ball	42.9	60.	243	170,000	41.5°	35.3
Neophilaenus (froghopper)	2.7	4.2	1.1	760	41°	36.6
Croton seed	3.5	8.5	4.6	2,000	41°	37.5
Flea beetle jump	1.6	2.93	0.543	310	40°	37.9
Tennis ball	65.	50	157.	220,000	41°	38.3
Vicia seed	2.7	9.0	4.1	1,600	38°	49.9
Spherobolus glebal mass	1.25	10.7	4.5	890	29°	61.7
Ruellia seed	2.2	12.0	4.9	1,800	35°	66.5
Table tennis ball	40	20	9.9	53,000	34.5°	75.6
Rabbit flea jump	0.5	4.0	0.3	130	30°	80.8
Dwarf mistletoe seed	2.6	24	10.3	4,200	31°	83.0
Hura seed	16.	70.	30.	75,000	28°	94.0
Pilobolus sporangium	0.3	20.	0.82	400	17°	98.0
Cornus canadensis pollen	0.04	7.5	0.024	20	4°	99.6
Ascobolus spore	0.12	30	0.30	240	12°	99.7
Auricularia spore	0.0083	2.0	0.00039	1.1	1°	99.7
Sordaria spore	0.04	30.	0.06	80	7°	99.96
Sphagnum spore	0.03	50.	0.0355	100	2°	99.99
Morus pollen	0.023	170	0.066	260	1°	99.997
Gibberella spore	0.01	35.	0.0046	23	1°	99.997

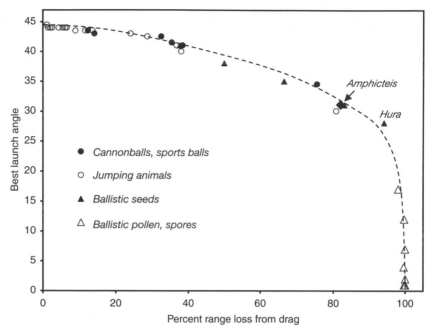

Figure 2.3. The relationship between loss of range to drag ("drag tax") and the launch angle above horizontal that yields the greatest horizontal range; launch speeds are given in table 2.1.

One might guess that a basketball, larger and less dense, would suffer relatively more from drag. But in practice its lower speed and thus relatively low drag (D, of course, not C_d) mitigates the problem. For a launch speed of 20 meters per second (72 kilometers per hour) it goes 35.7 instead of 40.7 meters, losing only 12.3 percent—but giving it an agreeably steepened descent as it travels basketward. The best launch angle drops only a little below 45°, to 43.5°. Tennis balls suffer more from drag. A good serve, at 50 meters per second would, if launched at the optimal angle of 41°, carry it 157 meters, but that distance comes after a loss of 38.3 percent to drag. Still more susceptible to drag is the very light ball used for table tennis. A slam at 20 meters per second would go less than 10 meters; best pitch angle has dropped to 34.5°, with a loss of three-fourths of its potential range to drag.

As drag becomes a greater factor, so too do both the dangers and opportunities for the player who puts some spin on the ball when hitting or throwing. Spin induces a force at a right angle to the direction of a ball's motion, "lift" whether downward, upward, or sideways, as a result of the Magnus effect. Sliced golf balls and tennis balls can deviate distressingly from their initial direction of flight; in table tennis much of the skill lies in deliberately sending off a rapidly spinning ball. But basketball players, no less fanatic, worry little about spin to alter trajectories.

Since so much in fluid mechanics depends on the Reynolds number, we might examine the present values for projectiles at launch. For the cannonballs, *Re*'s are about 5,000,000; for the basketball, 320,000; for the golf ball, 170,000; for the tennis ball, 220,000; and for the table tennis ball, 53,000. Clearly Reynolds number alone provides no easy key to the importance of drag. Nor does what we have called the drag tax depend exclusively on launch speed. We'll return in a few pages to the way one might predict the effects of drag—without an interactive program—on trajectories.

Where Drag Matters Little for Organisms ...

First, though, a look at existing data for biological projectiles, where necessary taking advantage of the computer to estimate launch speed from range and vice versa. Such data can be found for a wide variety of systems—the present account will be selective rather than exhaustive. Again, input data and results are summarized in table 2.1. To start with, consider a small, jumping mammal, a species of kangaroo rat (*Dipodomys spectabilis*) native to western North America and similar to the jerboa of North Africa and the marsupial kowari of Australia. It can be approximated as a sphere about 0.1 meters in diameter with a density of about 750 kilograms per cubic meter. According to Biewener et al. (1981), it can hop along bipedally at up to 3.1 meters per second (11.2 kilometers per hour), which implies a launch speed (above the horizontal, of course) of about 3.1/sin 45° or 4.4 meters per second (15.8 kilometers per hour). It achieves its best performance at a launch angle indistinguishable from 45° and incurs a drag tax of only about 1.1 percent. Why so little effect? Mainly, as we saw for the basketball, its decent size, fairly high mass, and low launch speed allow little opportunity for drag to exert its malevolence.

Among mammals that make haste with bipedal hopping, kangaroo rats are among the smallest. Simple consideration of surface-to-volume ratio—or, in effect, drag-to-gravity ratio—tells us that larger mammals will suffer even less from drag. So we anticipate that neither control of body posture, streamlining, nor altered piloerection will make much difference either to range, best launch angle, or speed. (By contrast, shape and postural changes do matter among animals that glide, where lift-to-drag ratio plays a crucial role, and among animals that "parachute," deliberately increasing drag to lower falling speeds.)

Similarly, drag should not be a significant factor for any fair-sized animal that locomotes with a sequence of short ballistic trajectories—one that goes arm-over-arm, brachiating from hand-hold to hand-hold (see, for instance, Usherwood and Bertram 2003). Nor will it matter for those amphibians that throw their prehensile tongues forward as prey-capturing devices, despite their impressive performances. For instance, the tongue of *Bufo marinus*, a large toad, accelerates at over 30 times gravity to launch at nearly 3 meters per second (10.8 kilometers per hour) (Nishikawa and Gans 1996); that of the salamander *Hydromantes imperialis* extends by 8 percent of its body length (Deban et al. 1997). Nor does drag make a great difference for a yet odder practitioner of ballistics. At least one insect lineage shoots fecal pellets, apparently to minimize their potential predator-directing role (Weiss 2003). The pellets of a skipper caterpillar, *Calpodes ethlius*, average 2.8 millimeters in diameter and about 930 kilograms per cubic meter in density. After launch at 1.6 meters per second (5.8 kilometers per hour) they go about 0.246 meters, only 5.4 percent below their dragless range, which they achieve at 1° below the dragless 45° angle (input data from Caveney et al. 1998). The pellets may be on the small side, but they don't go fast.

Smaller Jumpers

As one can see from table 2.1, as the size of jumpers drops, drag becomes increasingly important. But it can't be assumed important just because a jumper happens to be small. A tiny jumping spider (a salticid) does only a little worse than a small mammal, losing 1.4 percent of its dragless range (Parry and Brown 1959). Nor are the jumping larvae of a Mediterranean fruit fly (Maitland 1992), bristletails (Evans 1975), springtails (Brackenbury and Hunt 1993), or trap-jaw ants (Patek et al. 2006) much affected—all do best with 44° launches and follow the easy formulas of that first physics course; all have body lengths well below 10 millimeters. Trap-jaw ants, by the way, are reported to launch at an average angle of 27°, which takes them only about 82 percent as far as they would go at their higher, optimum angle. Because of their low launch speed, 1.7 meters per second, they lose no more to drag than the much larger desert locust (*Schistocerca gregaria*), about 6 to 7 percent of their dragless ranges (Bennet-Clark 1975; Patek et al. 2006).

But many small arthropods encounter real trouble. Data exist for five species of froghoppers (spittle bugs) (Burrows 2006) and at least one species each of click beetles (Evans 1972), flea beetles (Brackenbury and Wang 1995), and box moss mites (Wauthy et al. 1997). Their range losses run from about 6.5 percent to 38 percent, and their best launch angles run inverse with those numbers, dropping from 44° to 40°. The smaller the jumper, the slower must be its launch to avoid serious drag—or, put the other way, drag becomes particularly bothersome if a jumper is both small and speedy.

The extreme of small and speedy are fleas, and they run into greater relative drag than any other known jumping animals. According to Bennet-Clark and Lucey (1967), a rabbit flea (*Spilopsyllus cuniculatus*) about 0.5 millimeters in diameter takes flight at 4.0 meters per second (14.4 kilometers per hour). Drag reduces its range from 1.61 meters to a mere 0.3 meters, a loss of no less than 80.8 percent. And that best range (still assuming the game consists of long jumps across horizontal surfaces) happens with a launch angle of 30°. It lands at a speed no longer equal to launch speed but fully four times slower. (Bossard 2002 measured similar launch speeds for cat fleas.) So a flea's world is draggier and its trajectories less parabolic. Whatever their heading, fleas jump into the teeth of a sudden, severe windstorm.

EXPLOSIVELY LAUNCHED SEEDS

Plants and fungi obviously lack equipment for continuous propulsion. Still, dispersal of their propagules must be almost as important for them as locomotion is for animals. Many manage to give seeds, pollen, and spores high-speed launches, and they do so in ways that represent more biological and physical diversity and span a greater range of sizes and initial speeds than does jumping. Ballistic plants and fungi also make much greater use (with, again, lots of diversity) of elevated launch sites than do jumping animals. Nonetheless, the same physical imperatives apply. Drag gets relatively worse as size decreases, but so fast are some of these projectiles that drag can be a major factor even for fairly large ones—much as we saw for the newer cannon, golf balls, and tennis balls.

Among large ballistic seeds, the current champion appears to be a tropical tree, sometimes planted as an ornamental, *Hura crepitans* (Swaine and Beer 1977; Swaine et al. 1979). Its disk-shaped seeds (plate 2.1), sometimes used as wheels for children's toys, are about 16 millimeters across and 350 kilograms per cubic meter in density. They launch with quite an audible pop at prodigious speeds, as high as 70 meters per second (250 kilometers per hour). While taking horizontal range from ground level as benchmark gets a little artificial for a tree that grows to 60 meters, that speed can take them nearly 30 meters. Impressive as it sounds, that distance is but a small fraction of the 500 meters that a *Hura* seed would go in a vacuum—its range loss is 94 percent. Until recently, the launch speed of 70 meters per second was the fastest speed known in the plant kingdom, and it still represents the fastest for a macroscopic item. Coincidentally, it matches the maximum in the animals, the dive of a falcon (Tucker et al. 1998). A rationalization for its extreme performance will appear in the next chapter.

Smaller seeds that lift off at more modest speeds fall into the same pattern we saw in jumping insects—the smaller the draggier but only roughly. The 3.5-millimeter seeds of *Croton capitatus* (Euphorbiaceae), launched at 8.5 meter per second (30.6 kilometers per hour) and 41°, go 4.6 meters and pay a drag tax of 37.5 percent (Stamp and Lucas 1983). The 2.7-millimeter seeds of *Vicia sativa* (Fabaceae), launched at 9 meters per second (32 kilometers per hour) and 38°, go 4.1 meters and pay a drag tax of 49.9 percent (Garrison et al. 2000). The 2.6 millimeter seeds of the dwarf mistletoe, launched at 24 meters per second and 31°, go 10.3 meters, with a drag tax of 83 percent (Robinson and Geils 2006). The 2.2-millimeter seeds of *Ruellia brittoniana* (Acanthaceae), launched at 12 meters per second (43 kilometers per hour) and 35°, go 4.9 meters and pay a drag tax of 66.5 percent (Witztum and Schulgasser 1995a).

Explosive seed expulsion occurs less often in still smaller seeds, probably because the increased surface-to-volume ratio will result in a further increase in relative drag, whatever the specific aerodynamic regime. For that matter (and putting aside *Hura*) the ballistic seeds vary only minimally in size—from 2.2 to 3.5 millimeters in diameter. That limited range can be rationalized if, as seems to be the case, the ballistic seeds depend almost entirely on ballistics to get some distance from the parental plant. That is, wind plays at most a minor role. All lose roughly half of their range to drag; appreciably smaller and drag would impose too high a tax; appreciably larger and the number of seeds that can be produced would decrease. In addition, some ancestry-resistant compromise might be at work between the conflicting demands of ballistic versus wind-borne travel—in effect drag minimization versus drag maximization. For ballistics, large size, high density, and compact shape are preferable; for wind carriage, small size, low density, and ramose shapes work better. Perhaps spores and pollen are willing to deal with far greater drag where the parent organisms are too small to produce seed-sized propagules or where ballistic projection simply positions them better for wind dispersal. (Among others, Stamp and Lucas 1983, 1990 discuss such matters; other explosively discharged seeds they studied appear to fit in the same cluster.)

And Explosively Launched Spores and Pollen

Small size, though, has proven less discouraging for ballistic spore dispersal by fungi and an occasional ballistic moss such as *Sphagnum*. Most likely the short stature of most fungi reduces their ability just to liberate spores into air that's moving enough for effective passive travel. Conversely, with truly tiny propagules, even fairly dense spheres will have agreeably low terminal velocities in free fall, so shooting them even a short way upward may pay rewards in subsequent passive dispersal distances. The same arguments may apply to some cases of ballistic pollen. One example, the bunchberry dogwood, *Cornus canadensis*, flowers close to the ground; the white mulberry, *Morus*, though, is a proper tree.

The most famous fungal projectile, one from which a modern dance group takes its name, is the sporangium of the ascomycete, *Pilobolus*. *Pilobolus* erects its hypha (stalk) a few millimeters above piles of bovid and equid dung; the sporangium atop the hypha shoots off, along with a bit of cell sap, at an initial speed of 20 meters per second (72 kilometers per hour). A sporangium (of the density of water), 0.3 millimeters in diameter, should go 0.82 meters at a best angle of 17° after paying a drag tax of 98 percent. In fact, sporangia go two or three times that far, almost certainly because they carry that cell sap. It adds mass without much increase in diameter, and it may even provide a slightly streamlined tail. Early in its travel, when going fastest and thus covering the most territory, it has a Reynolds number, up to 400, high enough for such shaping to help.

Incidentally, that speed of 20 meters per second comes neither from my back calculation nor from stroboscopic pictures. Long ago, A. H. Reginald Buller (1934) adopted a technique first used (per his modest disclaimer) by Napoleon's technicians to measure bullet velocities. After firing through two coaxially rotating disks of paper they measured the offset of the second hole; that, together with the distance between the disks and their rotation rate, gave bullet speed. Buller put a perforated disk in front and an unperforated one behind, taking advantage of the sporangium's habits of shooting toward a light and sticking to whatever it hits.

Pilobolus probably makes little use of wind. It fires shortly after dawn, not a windy time of day, taking aim at the sun, at that time low in the sky. Perhaps it aims and times its launch to get that 17° angle that maximizes windless range—no one, I think, has looked into the matter. The objective is a bit of grass or other forage far enough from its own pat of dung to be attractive to another grazer—completion of its life cycle requires passing through a herbivore's gut, and large herbivores (parasite-privy consumers) prefer not to graze too closely to what we used to call horse-apples and cow-pies. I suspect that *Sphaerobolus*, a fungus that shoots a glebal mass a bit over a millimeter across (Buller 1933), also relies solely on ballistics for dispersal.

A still smaller projectile has a shorter range despite a higher launch speed. Another ascomycete fungus, one once favored by geneticists, *Sordaria*, shoots eight-spore clusters, about 40 micrometers across. Ingold and

Hadland (1959) give it a typical range of about 60 millimeters, from which I calculate an initial speed of 30 meters per second (108 kilometers per hour) and a drag tax of 99.96 percent. If horizontal range were the objective, its best launch angle would be a mere 7°—in fact, it seems to shoot upward. Why shoot at all? Further above a surface implies greater ambient air movement. With a terminal velocity below 50 millimeters per second, an upward shot would expose it to moving air for about a second, enough time for even a modest wind to provide significant lateral transport.

Data exist for a few other ballistic spores—the ballistic habit turns out to be remarkably widespread, with both the systematic positions of the shooters and the diversity of launching mechanisms indicating multiple evolutionary origins. The peat moss, *Sphagnum*, uniquely (as far as I know) squeezes air rather than water, shrinking the short circumference of an ellipsoidal reservoir into a tight cylinder. In this way it raises the air pressure to about 500,000 pascals (5 atmospheres), according to Ingold (1939) and Maier (1974). Far more compressible than water, air will store more energy at a given pressure, so it can serve as the elastic propulsive reservoir rather than, as with hydrostatic systems, transferring the task to the container's walls. From range and size I calculate the remarkable launch speed of 50 meters per second (180 kilometers per hour).

The fungus *Ascobolus* generates pressures almost as high. Its osmotic engine produces up to 300,000 pascals (3 atmospheres) to shoot, like *Sordaria*, an octet of spores for which Fischer et al. (2004) calculated a speed of 9.2 meters per second. Their figure, though, depends on Stokes' law for drag. During the critical first phase of the trajectory, though, the Reynolds number is 30, well above 1, meaning that drag will be much worse (eq. 2.6). Launch speed, then, must be higher than they assert in order to achieve the reported range. From that range I estimate 30 meters per second (108 kilometers per hour), the same as that for *Sordaria*.

Fungal guns come even smaller. Recently Frances Trail and Iffa Gaffoor measured a range of 4.6 millimeters for individually ejected ascospores of *Gibberella zeae*, a corn pathogen. Initial speed, a lot harder to measure, became important in justifying the high launching pressure and in identifying the responsible osmolytes, so I was drawn into the project. From spore density, about that of water, and size, about 10 micrometers, I calculated the remarkable launch speed of 35 meters per second (125 kilometers per hour). That speed, without drag, would take a spore 125 meters, so drag costs it no less than 99.997 percent of its potential range. It reaches its best range at a 1° launch angle. With a terminal velocity under 3 millimeters per second, it goes nearly as far (albeit briefly) upward from the launch site if shot vertically as it can go horizontally at its best launch angle (Trail et al. 2005). With a vertical shot from the ground it will be exposed to moving air for about a second and a half, far longer if launched from the surface of a plant that harbors the fungus. So in the real world even that low range should ensure excellent displacement, getting it out beyond most or all of the low-speed air near the launch site for a decent period of time.

Even smaller than a *Gibberella* spore is that of *Auricularia* (Pringle et al. 2005), 8.3 micrometers in diameter. With a slower launch—2.0 rather than 35 meters per second—it suffers less, relatively speaking, from drag, a mere 99.7 percent. With an initial Reynolds number just above 1.0 (and uniquely among the present projectiles), a calculation of range based on Stokes' law should be trustworthy. If smaller projectiles exist in nature, we know nothing about their behavior. And (as chapter 6 will elaborate), going much below 10 micrometers in air brings a different kind of physics into the picture. Along with the rest of ordinary fluid mechanics, Stokes' law assumes that fluids are continua. That assumption breaks down, not at molecular dimensions, but when the mean free paths of molecules—in gases of the order of 0.1 micrometers—become significant.

Nor is *Gibberella* the fastest known microprojectile; it held that record for only a few months. Taylor et al. (2006) claim speeds of 170 meters per second (over 600 kilometers per hour) and above for pollen grains of the white mulberry tree *Morus*! Popping their data for size and range into my computer program gives just the numbers they report, so I see no reason to doubt these astonishing speeds. Vascular plants now leave both fungi and animals in the dust. Fortunately (I suppose), these projectiles have both a limited range and insufficient momentum to make even minor mischief.

GENERALIZING

Why should a higher drag tax inevitably be associated with a lower optimal launch angle? The answer comes from the fact that with greater relative drag, descents are both steeper and slower than ascents. So the rule for maximizing range in a draggy world is simple—get one's distance while one still has decent speed, and don't waste a fine launch speed going in a minimally useful direction, in particular, upward. A look at some draggy trajectories that have been marked at uniform time intervals, as in figure 2.2, puts the matter with graphic directness.

Can aerodynamic lift be used to extend the range of a ballistic projectile? True airfoil-based gliding, used in many lineages of both animals and plants, requires a fairly specialized shape. Another possibility, though, consists of Magnus-effect lift—spinning or tumbling in such a way that the top of the projectile moves in the opposite direction of its overall flight and thus moves with (or at least less rapidly against) the oncoming air. The effect, mentioned earlier, goes by various names in our various sports—

slicing, top-spin, and so forth. But whatever the name and whether desired or counterproductive, it causes a projectile to deviate from the trajectory that gravity and drag would otherwise determine. Any gain, though, will be small, probably less than 10 percent in a large seed or jumper and quite a lot less in small forms. Springtails, small flightless insects (order Collembola), appear to use the device (Brackenbury and Hunt 1993), spinning with their upward surfaces moving with the wind at about 16 revolutions per second. Other suggestive cases await investigation.

How might an organism project tiny propagules with less severe limitations than those experienced by spores and pollen? Neither jet nor rocket propulsion occurs in aerial systems, but both the requirement for very high prelaunch accelerations (elaborated upon in the next chapter) and the disability imposed by drag (as here) can at least be ameliorated. The moss, *Sphagnum*, may do one or both, although I can't point to a specific investigation. *Sphagnum* makes a nearly spherical capsule atop a stalk that extends well above its green gametophyte body. When the lid blows off its capsule all the spores go out in the resulting blast of air (Ingold 1939). As a result of that brief tail-wind, they do not immediately encounter the full oncoming relative wind set by their speed. And they go off in a cloudlike group. That should permit drag reduction by what in our vehicular world we call "drafting" and which works far better in the very viscous regime of tiny particles—in effect, pooling mass and reducing effective surface area.

The present chapter, like its predecessor, is intentionally eclectic, deliberately bringing together material made heterogeneous by our traditional disciplinary divisions. Contriving effective comparisons all too often entails looking at how something performs under circumstances that may be adaptively irrelevant. Thus, as noted, *Pilobolus* should (and may) pick a launch angle that gives greatest still-air range, but the other spores and pollen almost certainly should not, inasmuch as they combine ballistics with wind dispersal. Seeds and skipper caterpillar pellets land at lower heights than those from which they were launched; *Hura* trees, for instance, grow quite tall. So real best ranges and optimal angles require further input data and adjustments of the basic computer program. But beyond exposing underlying commonalities, bringing disparate material together directs attention to gaps in understanding and to investigational opportunities.

PREDICTING AND MODELING

Finally, we might explore the utility of an index that indicates the degree to which drag will alter a ballistic trajectory—a "drag index." One initially suspects that Reynolds number (eq. 2.5) should do the job. After all, we've seen that many small projectiles suffer especially great losses to drag. But,

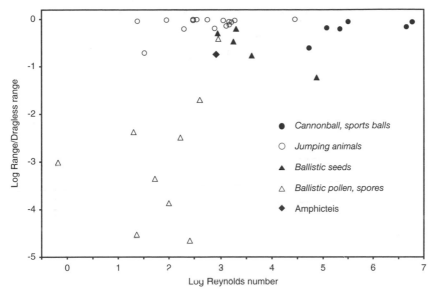

Figure 2.4. Reynolds number versus real maximum range relative to dragless maximum range. No strong predictions can rely on such scatter!

as in figure 2.4, it's demonstrably inadequate, especially when one notes the extreme range of the axes on the graph. If we limit ourselves to trajectories in air, the density and viscosity of air must be blameless. In the end, differences in speed must be the flies-in-the-ointment. A lot of small projectiles launch at remarkably high speeds, speeds never reached by larger ones.

I've found no index that gives a precise prescription, in part because drag cannot be reduced to some simple proportionality. Still, we can produce an index with little difficulty, one that does much better than the Reynolds number. Its basis is the ratio of the two forces that contribute to the form of a trajectory, drag and gravity. For drag we might use the product of pressure drag and viscous drag. Pressure drag is proportional to the product of density, speed squared, and length squared, $\rho_m v^2 l^2$, viscous drag to the product of viscosity, speed, and length, $\mu_m v l$, with the subscript "$_m$" referring to the fluid medium. (Adding the components of drag sounds better, but it won't work for proportionalities.) Gravitational force is proportional to mg or the product of density, length cubed, and gravitational acceleration, $\rho_\pi l^3 g$, with the subscript "$_p$" for the projectile. Squaring gravitational force keeps the index dimensionless; taking the square root of the ratio of the product of the two forms of drag to the square of gravitational force keeps values from getting unwieldy. (If you're dimensionless, so are your roots.) Thus we have

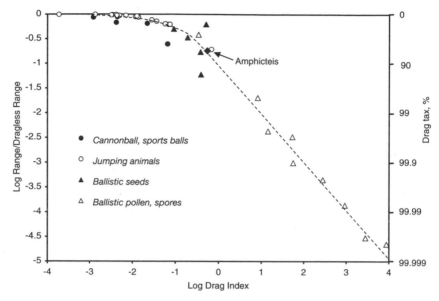

Figure 2.5. The relationship between the drag index (eq. 2.7) and real maximum range relative to dragless maximum range (left scale) and "drag tax" (right scale).

$$DI = \left[\frac{\rho_m \mu_m v_o^3}{\rho_p^2 l^3 g^2} \right]^{0.5} = \left[\frac{2.250 x 10^{-7} v_o^3}{\rho_p^2 l^3} \right]^{0.5}_{air} \tag{2.7}$$

where v_o is launch speed. The version on the right includes the SI values of the room temperature density and viscosity of air and of normal gravitational acceleration; it thus assumes SI values for the remaining variables[*].

This drag index suggests two things for biological projectiles, among which density varies by little more than a factor of two. First, low values, meaning minimal effects of drag, will characterize large objects traveling slowly—such as jumping mammals. Conversely, high values, meaning substantial drag effects, will occur with small objects going rapidly. Reynolds number, our usual index for the nature of a flow, includes the product of length and speed; this index uses their ratio. What about specific values of the index? Figure 2.5 plots index values for the cases discussed earlier against real range relative to dragless range. One sees that major effects of drag occur when projectiles are very fast (the *Hura* seed) or very small (the pollen and spores), although substantial effects (note the logarithmic scales) happen, as expected, for more ordinary items. Crude as it is, the

---

[*] The index is the inverse of the one in the original chapter; this one seems easier on the intuition.

drag index should help anticipate the performance of yet other biological projectiles without having recourse to a recondite computer program.

Second, for many fluid problems what matters is the ratio of viscosity to density, the so-called kinematic viscosity (as in the Reynolds number). Air and water differ only about fifteen-fold. By contrast, this drag index contains the product of viscosity and density in which air and water differ almost 50,000-fold. For that reason, shifting to water will cause the index to leap upward, and buoyancy will decrease effective g, raising it further. That rationalizes the scarcity of underwater ballistic devices in either nature or human technology. The examples often cited of aquatic guns turn out to be poison-injecting contact weapons, not ballistic propulsors—the nematocysts of *Hydra* (Holstein and Tardent 1984) and other Cnidaria and the radular tooth of the mollusk *Conus* (Shimek and Kohn 1981).

The word "scarcity" for the occurrence of aquatic jumpers and guns in the original version of this chapter turned out to be fortuitous; I might well have asserted "absence" in nature and been caught short. Peter Jumars subsequently told me about an aquatic case that he had investigated (Nowell et al. 1984), that of an infaunal polychaete worm, *Amphicteis scaphobranchiata*. The creature extends itself upward and tosses its fecal pellets out of its feeding pit. It throws the roughly 3-millimeter particles, a bit denser than seawater, at 0.27 meters per second. In vacuo, that would take them 7.4 millimeters, but that drops to 1.3 millimeters when one puts seawater values for viscosity and density into the program otherwise used for aerial ballistics. And the fecal pellets do go barely beyond the ends of the tentacles that toss them. Still, they're up and out of the feeding pit and may be carried some distance farther by local currents. Faster launches would gain little—one notes that the flealike 82 percent drag tax comes at a launch speed below that of any of our aerial cases. Incidentally, on the graphs of neither figure 2.4 nor 2.5 does *Amphicteis* occupy a particularly anomalous position.

The drag index also serves as a loose rule for making scale models. It enables a person to get a feel, through a bit of hands-on activity, for the world of very draggy projectiles. Just weigh a balloon, inflate it, and throw it as far as possible. Estimate launch speed from equation (2.2) and the range of a thrown projectile of minimal drag—perhaps a golf ball. From mass, size, and speed, you can then calculate a trajectory index. For a 150-millimeter, 0.66-gram balloon, I got a value midway between those of a jumping flea and a *Pilobolus* sporangium, suggesting a range loss around 95 percent—about what happens when I throw the balloon.

Dimensionless ratios have given long and valuable service as rules for making dynamically similar models. Sometimes, as in the case of the Froude number and ship-hull testing, that role has provided the initial impetus for a ratio's exploration.

Appendix

The program for incorporation of drag in calculating ballistic trajectories.

1. Enter diameter, density, initial speed, and pitch angle for the projectile, assumed to be spherical (with approximations if not).
2. Enter a time increment that will produce 500 to 2,000 iterations—adjust after experience.
3. Enter branching instruction for whether or not to include drag; if dragless, go to #7 below.
4. Include viscosity and density of the medium (change when needed) and gravitational acceleration.
5. Calculate volume and cross-sectional area of projectile.
6. Calculate Reynolds number (eq. 2.5) for projectile.
7. Estimate drag coefficient based on Reynolds number and cross-section (eq. 2.6 and text figures just above it). For dragless trajectories, $C_d=0$.
8. Calculate drag from drag coefficient ($D=0.5\ C_d\rho Sv^2$, where r is the density of the medium, not the projectile).
9. Calculate the decelerative effect of drag (drag over projectile mass), then break it up into vertical (multiply by sine of pitch angle) and horizontal (multiply by cosine of pitch angle) components.
10. Add the gravitational acceleration to the vertical component.
11. Calculate the resultant velocity after a time increment from these accelerations.
12. Calculate the coordinates of the resultant position from average of that and the earlier velocity and the new direction.
13. Increment time-increment counter, keep track of maximum y-coordinate.
14. Ask whether y-coordinate of new position is less than or equal to zero; if so, go on to report; if not, loop back to step #6.
15. Report: Horizontal range (final x-coordinate); maximum height (from #13), number of iterations, elapsed time, final speed (last #11).

CHAPTER 3

Getting Up to Speed

LEAPING, LAUNCHING, AND LURCHING

Generalizations in biology come hard, so we treasure any that cut through life's overwhelming diversity. In his famous essay, "On Being the Right Size," J.B.S. Haldane (1926) notes that jumping animals of whatever size should reach the same maximum height, an insight he attributes to Galileo. Other iconic figures make the same assertion about the size-independence of height—Giovanni Alphonso Borelli (1680), grandfather of biomechanics; D'Arcy Thompson (1942), godfather of biomechanics; and then A. V. Hill (1938), father-figure for muscle physiologists.

The basic reasoning is straightforward, at least if drag can be ignored. The force of a muscle varies with its cross-sectional area. The distance a muscle can shorten varies with its length. So the work a muscle can do will vary with the product of the two, in effect with its mass. All mammals have about the same mass of muscle relative to body mass, roughly 45 percent, and other jumping animals scatter only a little more widely. So the work available for a jump should be proportional to body mass. At the same time, the energy, mgh, required to achieve a given height, h, should also be proportional to body mass, m (gravitational acceleration, g, of course, stays constant). Thus h should be invariant.

Put in slightly different terms, launch speed, v_o, sets height for a projectile shot upward, and kinetic energy at launch is $\frac{1}{2} mv_0^2$. So the energy required to achieve a given launch velocity, like the work available, will be proportional to body mass. Either way, height should not depend on body mass.

As Borelli (1680), in the first great treatise on biomechanics, put it "if the weight and mass of a dog are a fiftieth of those of a horse [ ] the motive force of the dog would be a fiftieth of that of the horse. Therefore, if the other conditions are equal [ ], the dog will jump as far as the horse." ("Force" for Borelli meant something close to what we recognize as work or energy; "energy" in the modern physical sense was defined by Thomas Young in 1807.)

The last chapter focused on the behavior of ballistic projectiles after launch. This one fleshes out the story by looking at what happens prior to launch, how projectiles of diverse sizes and functions reach what turn out to be fairly similar launch speeds.

THE SCALING OF ACCELERATION

What does constant jump height imply about prelaunch acceleration? The smaller the creature, the shorter the distance over which it can accelerate to that standard launch speed and the higher its acceleration must be. At first glance that raises no problem for muscle-driven launches. If force, F, scales with length squared (muscle cross section); mass, m, scales with length cubed (muscle volume); and by Newton's second law $a = F/m$, acceleration, a, should scale inversely with body length, that is, $a \propto l^{-1}$. In short, small jumpers should naturally achieve higher accelerations. Consider two adept mammalian jumpers. A lesser galago (or bushbaby) with a leg extension of about 0.16 meters accelerates at 140 meters per second squared, while an antelope with a leg extension of 1.5 meters accelerates at only 16 meters per second squared (Bennet-Clark 1977). The comparison comes close to the predicted inverse proportionality—so far, so good.

How does acceleration scale when we look beyond such muscle-powered animal systems to include other projectiles such as those whose trajectories were examined in the previous chapter? Table 3.1 compares body (or projectile, for nonjumpers) size with data or estimates for prelaunch acceleration. But bear in mind its limitations. (1) Its selection of systems makes no great claim to be representative, although it does span the whole size range for which we have data. (2) For want of any ready alternative, the entries assume steady prelaunch acceleration. (3) Accelerations not reported in the literature have been calculated from launch speeds and estimates (from body proportions) of prelaunch travel distance. For mammals, with negligible drag, launch speeds come from jump heights; for smaller projectiles the computer program mentioned in the last chapter was used to work back from reported ranges.

Two things emerge from a casual look at the table. First are the extreme accelerations of small projectiles. The current record holder, a pollen grain of the white mulberry, *Morus alba*, accelerates at a truly cosmic (really, as we'll see, microcosmic) 9,600,000 meters per second squared (Taylor et al. 2006). A *Gibberella* spore achieves 8,500,000 meters per second squared (Trail et al. 2005), and several other distantly related fungal spores get into the six-figure range. These figures far exceed the acceleration of a rifle bullet, typically 500,000 meters per second squared. And second, the sheer diversity of systems represented raises the suspicion that extraordinary accelerations might not require extraordinary engines.

Logarithmic graphs do lovely service in suppressing scatter and the uncertainties introduced by rough estimates, especially where data span many orders of magnitude. Figure 3.1 gives such a log-log plot for the data just tabulated; a linear regression of the logarithms gives a slope of

TABLE 3.1
Projectile sizes and estimates of prelaunch accelerations for biological projectiles. Lengths in meters; accelerations in meters per second squared. To convert the latter to multiples of gravitational acceleration ("g's"), divide by 10.

Projectile		Length	Acceleration	Source
Gibberella zeae spore	f	0.00001	8,500,000	Trail et al. (2005)
Sordaria fimicola, spore	f	0.00002	1,100,000	Ingold and Hadland (1959)
Auricularia spore	f	0.00002	120,000	Pringle et al. (2005)
White mulberry, pollen	f	0.000023	9,600,000	Taylor et al. (2006)
Sordaria, 8-spore cluster	f	0.00004	1,100,000	Ingold and Hadland (1959)
Bunchberry dogwood pollen	f	0.00004*	24,000	Whitaker et al. (2007)
Ascobolus immersus, spore	f	0.00012	1,000,000*	Fischer et al. (2004)
Moss mite (*Zetorchestes*)	a	0.0002	3,400	Krisper (1990)
Pilobolus sporangium	f	0.0003	100,000	Buller (1909)
Rat flea	a	0.0005	2,000.	Bennet-Clark and Lucey (1967)
Box moss mite (*Indotritia*)	a	0.0008	1,200	Wauthy et al. (1998)
Sphaerobolus glebal mass	f	0.0012	46,000	Buller (1933)
Geranium molle	s	0.0016	8,100	Stamp and Lucas (1983)
Flea beetle (*Psylliodes*)	a	0.002	2,660	Brackenbury and Wang (1995)
Springtail	a	0.002	47	Brackenbury and Hunt (1993)
Geranium carolinarium	s	0.002	10,300	Stamp and Lucas (1983)
Viola striata	s	0.0021	7,800	Stamp and Lucas (1983)

(continued)

TABLE 3.1 (*continued*)

Projectile		Length	Acceleration	Source
Ruellia brittoniana	s	0.0023	10,000	Witztum and Schul gasser (1995a)
Vicia sativa	s	0.0027	7,500	Garrison et al. (2000)
Skipper butterfly frass	a	0.0028	180	Caveney et al. (1998)
Geranium maculatum	s	0.0029	7,600	Stamp and Lucas (1983)
Dwarf mistletoe (*Arceuthobium*)	s	0.0029	90,000	Robinson and Geils (2006)
Croton capitatus	s	0.0035	5,200	Garrison et al. (2000)
Flea beetle (*Altica*)	a	0.004	100	Brackenbury and Wang (1995)
Trapjaw ant	a	0.004	6,800	Patek et al. (2006)
Impatiens capensis	s	0.0051	1,650	Stamp and Lucas (1983)
Froghopper (*Neophilaenus*)	a	0.0054	4200	Burrows (2006)
Blepharis ciliaris	s	0.0055*	11,000	Witztum and Schul gasser (1995b)
Salticid spider	a	0.006	51	Parry and Brown (1959)
Desert locust, 1st instar	a	0.007	200.	Katz and Gosline (1993)
Froghopper (*Philaenus*)	a	0.0082	5400	Burrows (2006)
Fruit fly larva	a	0.0085	860	Maitland (1992)
Froghopper (*Lepyronia*)	a	0.0096	3067	Burrows (2006)
Click beetle	a	0.010	3,800	Evans (1972)
Bristletail (*Petrobius*)	a	0.010	400	Evans (1975)
Froghopper (*Aphrophora*)	a	0.0128	2267	Burrows (2006)
Froghopper (*Cercopsis*)	a	0.0128	2533	Burrows (2006)

(*continued*)

TABLE 3.1 (conitnued)

Projectile		Length	Acceleration	Source
Hura crepitans	s	0.016	41,000	Swain and Beer (1979)
Acris gryllus	h	0.027	64	Marsh and John-Alder (1994)
Pseudacris crucifer	h	0.029	58	Marsh and John-Alder (1994)
Desert locust adult	a	0.040	160.	Katz and Gosline (1993)
Hyla squirella	h	0.044	29	Marsh and John-Alder (1994)
Hyla cinerea	h	0.056	26	Marsh and John-Alder (1994)
Jumping mouse	m	0.07	143	Nowak (1991)
Anolis carolinensis	h	0.07	45	Toro et al. (2003)
Osteopilus septentrionalis	h	0.088	26	Marsh and John-Alder (1994)
Jerboa, kowari, kangaroo rat	m	0.12	75	Nowak (1991)
Red squirrel	m	0.15	60	Essner (2002)
Lesser galago	m	0.16	140.	Bennet-Clark (1977)
Rana catesbiana	h	0.164	20	Marsh (1994)
Potoroo	m	0.4	100	Nowak (1991)
Springbok	m	0.8	125	Nowak (1991)
Impala	m	1.0	100	(various)
Cougar (mountain lion)	m	1.0	55	Nowak (1991)
Gray kangaroo	m	1.1	67	Nowak (1991)
Horse, eland	m	1.7	80	Nowak (1991)

Notes: (f: spore or pollen grain; s: seed; a: arthropod; h: frog or lizard; m: mammal.) Asterisks indicate data that I've recalculated or obtained from a secondary source.

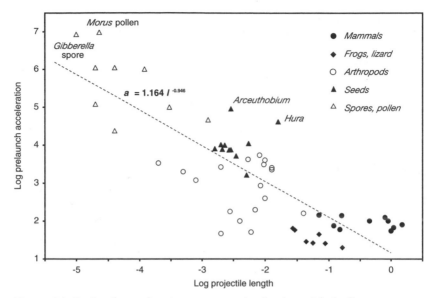

Figure 3.1. Projectile acceleration versus projectile size, with the linear regression line and equation for the data set. $r^2 = 0.68$.

−0.946. That scaling exponent comes reasonably close to the value of −1.0 corresponding to constant launch speed, especially when one considers the diversity of both organisms and engines. In statistical terms, the exponent does not differ from −1.0. In a sense, the fit comes as something of a surprise since no effort was made to exclude bad jumpers, especially among the frogs and insects. Including rhinos, polar bears, and other unlikely leapers among the big mammals would have offset the inclusion of some underperforming smaller animals and pushed the exponent still closer to −1.0. Our forefathers have been vindicated—asserting that all creatures can jump to the same height implies just the scaling relationship for acceleration that we indeed find.

The figure holds other messages as well. For a start, the scatter ought to be examined. Seeds in general do better than arthropods of about the same size. A *t*-test supports that impression, yielding a significant difference between the size-acceleration products for the two groups (excluding the seeds of *Arceuthobium* and *Hura*, the extreme outliers). Where sizes overlap, both arthropods and mammals do better than frogs. Despite Mark Twain's famous short story, frogs (although no tested specimens seem to have been collected in Calaveras County) don't jump all that well—they're just sedentary creatures from which single long jumps can be elicited easily.

The surprisingly limited size range of the seeds (once again omitting *Arceuthobium* and *Hura*) relative to the other groups was noted in the last chapter. The suggested pair of constraints at work consisted of excessive drag loss for smaller ones and excessive reduction of seed number for larger ones, with the first constraint an issue where dispersal depended solely on ballistics, with no contribution from wind.

The Limitations of Real Engines

The diversity of cases for which the scaling rule works ought to raise a flag of suspicion. Why should an argument based on muscle work for systems that do their work with other engines? Muscle represents no typical biological engine—it ranks at or near the top in, for instance, power-to-mass ratio. For that reason, if nothing else, we cannot conclude that the inverse proportionality confirms our reasoning. Something else must be afoot—again, the original argument presumed isometrically built muscle-powered jumpers. In short, the fit of the far more diverse projectiles demands a more general argument for the scaling of projectile performance. The way the original rationale brushes aside the sometimes extreme effects of aerodynamic drag received attention in the preceding chapter. In addition, the rationale sweeps aside a serious biological limitation, a matter of how muscles really perform. It also sweeps aside a major physical presumption, one that once examined, cannot be dismissed.

The biological limitation comes from the way muscle performance scales—or fails to scale. The work a muscle can do, relative to its mass, depends little on its size or on that of its animal. But consider jumpers impelled upward by muscles that shorten as they jump—shortening, as one might say, in real time. An invariant launch speed demands that the muscles of the smaller animal do their work in a shorter time.

Skeletal muscle differs only a little either from muscle type to muscle type or with animal size, at least in our broad-brush view. The resting length of the basic contractile unit, the sarcomere, is about 2.5 micrometers in vertebrates, and it varies by less than an order of magnitude elsewhere if we exclude a few odd extremes. Muscles consist of sarcomeres in series, so if all sarcomeres shorten at the same speed, then contraction speed should be directly proportional to muscle length. Or, put another way, speed relative to length, called intrinsic speed and given in units of inverse seconds, should not vary with body size. (Intrinsic speed, not a speed in the strict length-per-time sense, equals minus the "strain rate," as usually used in the engineering literature; the change of sign reflects a shift from shortening during normal muscle action to stretch during a tensile test.) The shorter the muscle, the slower its contraction speed at a given intrinsic speed.

Muscle does not operate with equal effectiveness over a wide range of intrinsic speeds, so an individual muscle does not operate with equal effectiveness over a wide range of actual speeds. A muscle pulls most forcefully (ignoring pulling during imposed extension) when not shortening at all, at zero speed. Force then drops off with speed until it hits zero at some maximum intrinsic speed. Power, force times speed, peaks at about a third of that maximum intrinsic speed. (McMahon 1984 gives a particularly good discussion of such matters.) In short, both force and power peak at speeds well below maximum.

Making a small animal power its jump by real-time muscle contraction until it reaches the same peak speed as does a large animal demands that its muscles operate at higher intrinsic speeds. Such speeds either reduce effectiveness or else cannot be reached at all. For example, consider two animals with launch speeds of 5 meters per second. One is 1.2 meters high and has to get up to launch speed in a third of its height, or 0.4 meters. Working backwards from launch speed and distance gives an acceleration time of 0.16 seconds (and, incidentally, an acceleration of 31 meters per second squared). Say it jumps with muscles 0.3 meters long that shorten by 20 percent of their length in the process, or 0.06 meters. Thus its muscles shorten at a speed of 0.06 meters/0.16 seconds or 0.375 meters per second. Dividing by muscle length gives 1.25 inverse seconds as intrinsic speed, low enough to get a fine power output, perhaps a decent fraction of the 250 watts per kilogram that approximates muscle's practical maximum.

Contrast that with a similar animal a hundred times shorter, 12 millimeters. It must get up to speed in 1.6 milliseconds (with an acceleration of 3100 meters per second squared). Its 3-millimeter muscles must shorten by 0.6 millimeters, thus at an identical speed of 0.375 meters per second. But real-time muscle contraction of its shorter muscles takes a hundredfold higher intrinsic speed, 125 inverse seconds, well beyond what vertebrate striated muscle can do. I think a mouse finger extender holds the upper record, 22 inverse seconds, but biological systems have difficulty getting reasonable (if still suboptimal) outputs above about 10 inverse seconds; where power matters, 5 inverse seconds is hard to exceed.

Therefore the old argument that all animals can jump to the same height cannot be correct if based on real-time muscle work—the biological presumption mentioned earlier. At best the rationale might work above the body length at which necessary intrinsic speed becomes limiting. Jumping ability (in terms of height) ought to drop off for animals less than 50 to 100 millimeters long but ought to be constant above that size. Cougar and kangaroo have muscles that yield more work per contraction relative

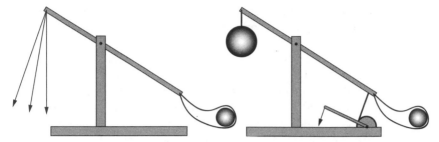

Figure 3.2. Diagrammatic versions of two large types of artillery powered by human muscle; both throw their projectiles from slings. The traction trebuchet, on the left, stored no energy except perhaps as kinetic energy of the moving arm; the counterweight trebuchet, on the right, made heavy use of gravitational storage.

to their masses and can jump higher than can jerboa and kangaroo rat. Drag is not the culprit (as noted in the last chapter); they do indeed have higher launch speeds.

Amplifying Power

The physical presumption underlying the old argument has jumpers conserving power as well as work or energy—that is, it asks muscle to power jumps in real time. In fact, most of the smaller jumpers evade the limitation on muscular performance just discussed by using power amplifiers to reach their necessarily higher accelerations. After all, conservation of work or energy does not imply conservation of power for nonsustained tasks. A system need only put energy in slowly and then release it rapidly—as done in archery.

A look at some large, muscle-powered weapons provides a direct comparison between devices lacking and ones equipped with power amplifiers. Prior to the advent of cannon, Medieval Europe and Asia attacked fortifications with first one and then another version of a catapult, the two devices called, respectively, traction trebuchets and counterweight trebuchets (Hill 1973), as shown in figure 3.2. A traction trebuchet applied power in real time—the artillerymen pulled simultaneously downward on one arm, raising the arm with its projectile-bearing sling on the other side of the fulcrum. A counterweight trebuchet stored energy gravitationally—artillerymen pulled (or cranked) the arm downward with the sling and projectile, slowly raising a mass of as much as 10,000 kilograms on the other end. Releasing a catch on the lowered arm allowed the counterweight to plummet, raising that arm with its sling and projectile.

Combining historical information with a few assumptions, I once tried to calculate the performances of representative trebuchets of the two kinds (Vogel 2001).

A traction trebuchet could throw a 60-kilogram mass a distance of 90 meters, implying a launch speed of 30 meters per second. A human can pull downward for a distance of a meter with a force of 220 newtons, doing 220 joules of work per pull. Since the projectile needs $1/2 \times 60 \times 30^2$ or 27,000 joules per shot, at least 120 artillerymen had to pull—assuming massless weapon arms and other unlikely idealizations. At Sind, now in Pakistan, in 708 CE, 500 people reportedly worked a single weapon—a yield of about 50 joules per operator per shot.

A counterweight trebuchet could throw a 225-kilogram mass a distance of 260 meters and thus had a launch speed of 50 meters per second and an energy of 300,000 joules per shot. Without power amplification by gravitational energy storage, that would have demanded not 120 but about 1400 artillerymen, again assuming perfect efficiency. As best we can tell, only about 50 were so employed—an effective output of 6000 joules per operator per shot, over a hundred times greater.

In the anatomically simplest power amplifier, muscles preload some elastic component. In that way a jump can be powered by the combined energy of direct muscular action and of elastic recoil of energy put in earlier by antagonistic muscles. Most or all of the vertebrates—and certainly all the smaller ones—listed in table 3.1 augment direct, real-time muscular action with some preloading of elastic components. The main storage sites are the tendons in series with the jumping muscles themselves. These complex muscle-tendon systems have yet to be fully analyzed, but they appear to involve initial crouching countermovements and some kind of catapult mechanism—at the least (Alexander 1988a; Aerts 1998).

Calculations of the power outputs of jumping muscles often give values well above what isolated muscle can do. In the absence of more specialized devices, that points to just such preloading. A tree frog, *Osteopilus*, for instance, achieves a peak output in a jump about seven times higher than the maximum output of the muscles it uses (Peplowski and Marsh 1997). It may be extreme, but its power booster is not unique among frogs. With such amplification, frogs keep their intrinsic speeds fairly low, below about 5 inverse seconds (Marsh 1994). A jumping bushbaby (*Galago senegalensis*) does something similar. Aerts (1998) calculated that for direct action of leg extensors to power its jumps, those muscles would need to weigh twice as much as the entire body. With big hind limbs, frogs and bushbabies are clearly specialized as jumpers. But little obvious structure underlies their power amplifiers—they seem to do their tricks much the way an eager (and abusive) automobile driver races the engine before engaging the clutch or automatic transmission.

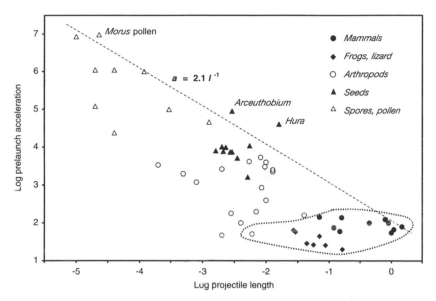

Figure 3.3. The same data and graphing conventions as in figure 3.1, but now showing which organisms make minimal use of elastic energy storage (enclosed) and with a line representing equivalent performance (a slope of −1) instead of the regression line.

Structures for power amplification appear almost universal among jumping arthropods—only one group of jumpers seem to lack any specialized device. Despite the name, the salticid spiders do not jump especially well, at least by the criteria of launch speed and acceleration. Parry and Brown (1959) looked hard but found no amplifier. Spiders, though, are something of a law unto themselves, since they extend their legs using hydraulics rather than with direct action of limb muscles.

Jumping with real-time muscle action or with simple preloading of tendon has its limits. Figure 3.3 fleshes out the picture presented by figure 3.1. A dotted ring encloses data for animals—frogs, lizards, mammals, and salticid spiders—that use minimal or no power amplification. For accelerations above about 150 meters per second squared, nature does not use real-time muscle action, even in such mildly augmented form, as the main impetus, whatever the size of creature or projectile. So we're left with a regression line in figure 3.1 with a slope (−0.946) statistically indistinguishable from the value predicted (−1.0) on the basis of an argument now revealed as specious, at least for muscle-powered jumpers.

How to Really Amplify Power

So far we've focused on muscle as engine. Again, both muscle's power rela-tive to weight and speed relative to size put it near the high end among living engines. Direct or largely direct powering of ballistic launches may be possible for muscular systems, at least for large ones. By contrast, where lesser engines drive launches, power amplification must be an abso-lute necessity for any kind of ballistic travel.

What, then, are the options for serious amplification? Linking a slow in-put with a rapid output requires a way to store energy. Our human technol-ogy employs such things as flywheels and rechargeable chemo-electric batteries, schemes with only distant analogs in nature. Both human and natural technologies use gravitational storage—from counterweight trebu-chets to pendulums in the former; stride-to-stride energy storage in legged walkers in the latter. One can imagine trees that toss fruit from wind-driven branches that sway as gravitational pendulums or seeds propelled by the drop of an elevated column of liquid, but I know no specific case of gravita-tional storage in biological ballistics. Most prelaunch amplifiers depend on another scheme, energy storage in deformed elastic materials. One kind of organism, though, stands as at least a semi-exception.

The jelly fungi, *Auricularia and Itersonilia* (figure 3.4a), osmotically produce droplets of cell sap at the bases of the spores they will launch. The droplet ("Buller's drop" after Reginald Buller, who first described it) col-lapses over the spore at launch. Apparently the energy released by relief of surface tension, via the change in the center of mass of the system, provides the immediate power source. So it creates surface, which takes energy, rather like stretching, compressing, bending, or twisting a conventional elastic material (Pringle et al. 2005). The dimension of surface tension, force per distance, translates directly into energy per area—as opposed to energy per volume for the input and output of a material elastic.

Remind yourself of the simplicity and ubiquity of power amplification through brief elastic energy storage by flipping the nearest toggle switch, one that controls the room lights or some piece of household electronics. With most such switches, one pushes a lever with increasing force until it abruptly stops opposing your effort and switches to its alternative posi-tion. You have slowly loaded a spring, which then rapidly releases that energy so the switch makes a sudden and robust change of electrical con-tact. You may continue to push, but you do little additional work once the spring has shifted from absorbing to releasing energy.

A single-shot amplifier, as in most ballistic plants and fungi, can be self-destructive and thus even simpler than a spring-assisted switch. The fungus *Pilobolus*, for instance, bears its sporangium atop a liquid-filled

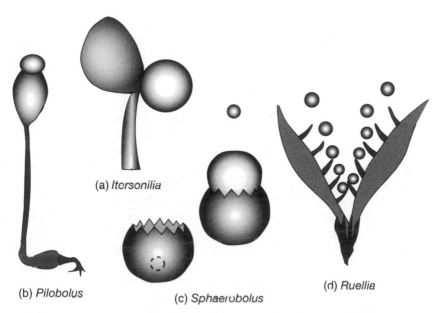

(a) *Itersonilia*

(b) *Pilobolus*

(c) *Sphaerobolus*

(d) *Ruellia*

Figure 3.4. The diverse launching devices of three ballistic organisms. (a) the fungus *Itersonilia*, with "Fuller's drop" to the right of the spore, just before lauch. (b) *Pilobolus*, shown with the sporangium on top of the subsporangial swelling just before it shoots upward on a jet of cell sap. (c) *Sphaerobolus* just before and just after a glebal mass of spores gets sent aloft by eversion of the floor of the cup. (d) the seed *Ruellia* just before the end of launch, with each seed propelled upward by motion of the jaculator beneath it.

hyphal tube, as in figure 3.4b. An osmotic engine raises the pressure in the tube until the sporangium suddenly detaches along a specific junctional line and takes flight (Buller 1909). That commonest of fungal schemes gets tweaked by ones such as *Sordaria* that manage to avoid self-destruction long enough to loose a series of up to eight spores in quick succession (Ingold and Hadland 1959). Another fungus, *Sphaerobolus*, uses a one-shot catapult in which an initially concave cup ("peridium") containing a millimeter-wide glebal mass of spores suddenly everts, becoming convex upward (Buller 1933; figure 3.4c). A similar bistable system has recently been described by Forterre et al. (2005) in a higher plant, the Venus's flytrap. In both fungi and flytrap, the ultimate engine is osmotic, coupled hydraulically with the output device.

Many seed shooters use another single-shot system, one in which drying of an initially hydrated structure such as a seed pod gradually stresses some woody (cellulosic) material. The movement accompanying breakage then sends the seed (or a group of seeds) onward. In *Ruellia*, for instance (figure 3.4d), sudden lengthwise rupture of the seam between two external valves (each analogous to a half-shell of a bivalve mollusk) lets the valves bend outward. That causes arms attached to the insides of valves to bend upward, which in turn causes each arm to push a seed up and out (Witztum and Schulgasser 1995a).

Among non-self-destructive amplifiers, that of locusts and grasshoppers is especially straightforward. According to Bennet-Clark (1975), rather than powering a jump directly, the large extensor muscle of each hind tibia loads a pair of elastic elements. A catch near the junction of the (proximal) femur and (distal) tibia keeps the leg flexed—a jump must start with fully flexed tibias. Relaxation of the flexor muscle releases the catch, and the immediate power for the jump comes mainly from energy stored in the chitin of elastic cuticle. Besides power amplification, the arrangement permits locust jumping muscle to operate at an efficiently low intrinsic speed, less than 2 inverse seconds. The peak power output of 0.75 watts from 70 milligrams of muscle represents almost 11,000 watts per kilogram, around 40 times what muscle can do directly.

Fleas (Bennet-Clark and Lucey 1967) have a more intricate mechanism, about which I'll give even less detail. In about 100 milliseconds the large trochanteral depressor muscles of the hind legs of a rabbit flea deform a pair of elastic pads, here not chitin but the softer and spectacularly resilient protein resilin. A second pair of muscles trigger energy release by moving a strap sideways, undoing the catch. The jump itself lasts only 0.7 milliseconds, quicker by nearly 150 times. Energy storage permits the muscles to operate at a very forceful intrinsic speed of 0.55 inverse seconds rather than an impossible 50 inverse seconds or more.

STORING ENERGY ELASTICALLY

Table 3.2 compares the key properties of several of the materials available for elastic energy storage by organisms, with spring steel included for reference. Of course ancestry constrains the choice of energy storage material. Thus only arthropods make resilin, and cellulose occurs mainly in plants. Furthermore, the storage materials of most projectile-producers represent only mild modifications of those of non-projectile-producing forebears. For resilience, work regained relative to work put in, resilin, known best from insect wing hinges, beats any other biological material. It may have to be superb, not because a few percent gain in resilience matters much in and of itself, but because the loss relative to perfect resilience (1.0) appears as heat. That may be nontrivial in insect wing hinges, where it may be alternately stressed and released hundreds of

TABLE 3.2
The relevant properties of materials for brief elastic energy storage and release (Bennet-Clark 1975; Gosline et al. 2002; Jensen and Weis-Fogh 1962; Vogel 2003). The numbers presume an uncomfortably large number of assumptions about such things as operating conditions and ignore large elements of biological variability.

Material	Energy/volume $(MJ\,m^{-3})$	Energy/mass $(J\,kg^{-1})$	Resilience	Relative elastic mass (%)
Arthropod cuticle	9.6	8,000.	~0.8	~0.2
Tendon (collagen)	2.8	2,500.	0.93	0.54
Wood	0.5	900.	~0.5	~2.3
Resilin	1.5	1,250.	0.96	1.04
Spring steel	1.0	150.	0.99	8.42
Air	0.000500	417.	1.00	0.75
Liquid water	0.18	180.	1.00	6.94

times each second. Fleas just happen to be members of an auspicious lineage.

Tendon is mainly collagen, our main elastic energy store (the protein elastin plays a lesser role); it also does well but has quite a different character and has to be used differently. Resilin, softer (elastic modulus 1.9 meganewtons per square meter), can be usefully loaded in compression or shear. Collagen, stiffer (1500 meganewtons per square meter), works best in tensile applications. Even there it can't be arranged like a conventional rubber—it operates at high forces and low extensions, crucial when linking muscles to bones where a stretchy string would undo the shortening of a contracting muscle.

Wood, despite our long use of wooden bows for archery, has poor resilience—we take advantage of its internal damping to make mellifluous resonators in many traditional musical instruments. But woods vary widely (as instrument makers have long known), and fresh woods are quite different from dry stuff. We know very little about the storage capabilities and resilience of the woods (and other cellulosic materials) that accumulate and release energy in nature's ballistic devices. On the one hand, ballistic seeds do impressively well, suggesting high resilience. On the other hand, the structures involved represent a tiny fraction of the total mass of a plant and probably an even smaller fraction of its lifetime energy expenditure, so efficiency relative to either mass or work may not be much of a factor. We know still less about the properties of the materials that fungi use for energy storage.

Air makes a perfectly fine elastic material, taking loads in either compression, almost without limit, or (pseudo-) tension, at least to 101,000 newtons per square meter, one atmosphere. I know of no case of its use in any ballistic system in nature, although I suspect that a moss, *Sphagnum*, might store energy for spore ejection by compressing air, as noted in the last chapter. Water, by its ubiquity and the data in the table, looks attractive; but that turns out to be illusory. Its extremely low compressibility (or high bulk modulus, the same thing) produces an awkward mismatch with biological solids. Squeeze water in a container of any such solid and the container stretches more than the water compresses—water requires operation at extremely high force with extremely low volume change. So, while water makes a superb medium for transmitting hydrostatic pressures, it's next to useless for storing energy.

The final column of table 3.2 gives a severely idealized calculation of the minimum mass of elastic material relative to the weight of the projectile needed for launch (so lower is better). In essence, it equates the initial kinetic energy of the projectile with the product of (1) the work (energy) of extension relative to mass of the elastic, (2) the resilience of the elastic, and (3) the mass of the elastic. The assumed launch speed of 5 meters per second corresponds to a dragless vertical ascent of 1.25 meters—a typical value for the present systems. On this perhaps biased basis the biological materials look remarkably good.

What Does Limit Acceleration and Launch Speed?

The old argument has crashed and burned. The work relative to mass of a contracting muscle deteriorates as animals get smaller rather than holding constant—a consequence of the requisite rise in intrinsic speed. Muscle need not and commonly does not power jumps in real time—elastic energy storage in tendons of collagen, in apodemes of chitin, and in pads of resilin provides power amplification. Finally, muscle powers none of those seeds and tiny fungal projectiles. Yet acceleration persists in scaling as the classic argument anticipates.

A look at the properties of elastic materials dispels any notion that their ability to store energy imposes a particular limit. Even the extreme case, launching a *Hura* seed with the energy of stretched or squeezed wood, would take an elastic mass only five or so times the mass of the projectile. That volume of elastic should be no problem, at least for shooters rather than jumpers. Distance (and thus speed) amplifying levers can compensate for inadequate speed of recoil of an elastic. And nature could probably enlarge muscular systems or run osmotic engines at higher pressures (although Alexander 2000 gives an argument against the first of these).

A possible alternative emerges if we reexamine the relationship between force and acceleration defined by Newton's second law. If acceleration indeed scales inversely with length and mass directly with the cube of length, then force should scale with the square of length. Or, put another way, force divided by the square of length should remain constant. Force over the square of length corresponds to stress, so we're saying that stress should be constant. Perhaps our empirical finding that acceleration varies inversely with length tells us that stress in some manner limits these systems. A design scheme that reflects some stress limit would represent no great biological novelty, having been recognized (or invoked) in remodeling of bone, in resizing of blood vessels, and in the growth of trees (see, for references, Vogel 2003).

The stress limit may go well beyond the maximum pull of a muscle. It might reflect a point of self-destruction, a limit that the propulsive equipment of a system might exceed only at risk, one might say, to life and limb. That could apply even to the largest jumpers, since experimental work on humans—anticipating rocket launches—shows that our bodies do not take kindly to accelerations much above those experienced by large mammalian jumpers. It also rationalizes the greater accelerations of seeds than of arthropods—seeds, simpler and sturdier, should be less easily damaged by high launch accelerations. I have to pick up small insects carefully lest I damage them; seeds we grind in the kitchen with mortar and pestle. (Of course seeds are not self-propelled fliers, the basis of an alternative or complementary explanation for their greater sturdiness.)

For stress limitation to provide a general explanation, the available materials must not vary inordinately in the maximum stress they can withstand— in their strengths, to use the usual terminology. That does less violence to reality than one might think. Sweeping aside a host of details and qualifying conditions, the strength of chitin (cuticle) and collagen (tendon) is about 100 megapascals, while that of keratin (hair and horn) and fresh wood is about 200 megapascals.

Perhaps we see in this scaling of acceleration a practical constraint imposed by the materials and structures of biological projectiles that must not be rendered dysfunctional by their ballistic episodes—these are whole animals and propagules, not bullets. That limit, emerging from the scaling of force with the square of length, might simply reflect the relative constancy and size-independence of the maximum stress tolerated by biological materials. It's consistent with (and may reduce to an example of) a more general scaling rule. Marden and Allen (2002) found just such scaling in the force output of a wide range of engines, ranging from molecular motors of myosin, kinesin, dynein, and RNA polymerase, through muscles to winches and rockets—their "group 1 motors." They attribute it to a common limit on just this capacity to withstand mechanical stress.

EXTREME CASES

Figure 3.3 replaces the regression line with a limit line of slope −1 over the graph's five orders of magnitude of size. If the basic scaling rule has acceleration running inversely with projectile length, then any line of that slope connects loci of what we might view as equivalent performances. This particular line takes a jumping cougar, the best of us mammals, as benchmark. It draws attention to several overperformers—the pollen launcher of the white mulberry, *Morus alba*, and the seed launchers of the dwarf mistletoe *Arceuthobium* and the tropical tree *Hura crepitans*. About the mulberry I can think of no special reason for why it might want to do so well. The performances of the two seeds, though, may have something to tell us.

One might guess that a seed needs to distance itself from some inhibitory parental root secretion (as is well known in some plants) or some pathogen with limited dispersal facility. I think, with the bias of the biomechanic, in particular one who focused on fluids, that something else might be at work. The dwarf mistletoe is a destructively parasitic plant that lives high in trees (never mind its secondary role as seasonal osculatory excuser). *Hura* is a tree, and a fairly tall one at that. All the other ballistically launched seeds for which I can find data are shrubs.

Seeds produced on or by trees need to go further than those of shrubs to get out from under the host or maternal shade. One might expect that higher launch points would solve the problem, but that expectation reflects our drag-ignorant bias. A high drag trajectory has a very steep descending portion, whatever its launch angle. Thus a *Hura* seed launched from 20 meters up at 28° has gone 29 meters when it descends to launch height; descending to the ground yields only an additional 3.5 meters—its downward course has an 80° average angle. (With a now-optimal 16° launch from the 20-meter branch, it could gain only another half-meter.) So escaping the maternal or step-maternal apron-strings must rely mainly on either larger size (lower drag relative to weight) or higher launch speed or both. Larger size extracts costs in terms of additional material needed to resist greater recoil (which sacrifices launch speed) and of fewer seeds for a given investment in mass. Higher launch speeds mean higher drag as well as greater recoil—*Hura* suffers an atypical 94 percent range loss, as noted in the previous chapter. Still, no matter how great the drag increase, speed increase always buys some increase in range.

At the same time, the present explanation of the inverse scaling rule depends on no inviolate basis for performance, one that makes exceptional cases difficult to rationalize. Stress is force per cross-sectional area; one need only provide greater area to allow greater force. Nothing really rules out a greater-than-typical investment in material—if circumstances warrant it. So one just needs to explain what's special about the circumstances.

Several final notes. The present chapter, following its predecessor, has focused on projectiles. Other biological systems achieve high accelerations, and these accelerations also vary inversely with size, despite their diversity of propulsive engines. So, for completeness, I ought to mention the ejectable nematocysts of the coelenterates, the retractable spasmoneme of vorticellid protozoa, tentacle extension in squid, jaw protrusion in sling-jaw wrasses, the protrusible tongues of many amphibians and reptiles, and ballistic jaw predation in trap-jaw ants.

And then we return to that assertion about all animals jumping to the same height. J.B.S. Haldane attributed it to Galileo; I believe he erred. I can find no such assertion or anything closer to it than his comment on the scaling of bones. I confirm D'Arcy Thompson's attribution to Borelli, down to chapter and verse. Borelli was only translated into English long after Thompson wrote *On Growth and Form*; but, as an accomplished classical scholar, Thompson would have read Borelli in the original Latin.

CHAPTER 4

Moving Heat Around

WHY MOVE HEAT?

We care about temperature. All too often we feel either too hot or too cold. Our appliances come with thermostats, cooling fans, and thermal protection switches. We determine the temperatures of organisms with thermocouples, thermal imaging equipment, and all manner of other thermometers. Temperature anomalies signal trouble, from personal fevers to global climate change. But the diverse and complex physical phenomena underlying temperature pose perilous pitfalls when it comes to explaining our data. Furthermore, we're easily misled by the intuitive sense that grows out of the experience of a large, terrestrial animal given to maintaining a steady body temperature close to its environmental maximum. We too easily forget that net photosynthetic rates for plants commonly peak at lower temperatures and that some of the most productive marine waters are quite cold.

In this and the next chapter, I want to look at the complexities of temperature and heat, asking which of the operative physical phenomena matter most, what options are open to organisms, what devices organisms use, and what as yet undemonstrated arrangements might yet be uncovered.

In few terrestrial habitats do organisms lack some thermal challenge. Where I live, in southeastern North America, temperatures range from about −19° to 38°C, on the old Fahrenheit scale a variation of no less than 100°. We intercept solar radiation of up to 1000 watts per square meter, and air movement can range from imperceptible to overwhelming. Breathing, a convective process, comes with the inevitable heat transfer of evaporation. Our own heat production adds an additional complication—a resting human generates about 80 watts; were an adult human to retain that energy, body temperature would rise by about a degree per hour.

Why not just accept the body temperature determined by the local interplay of such phenomena? As is so often the case, we can, at best, make educated guesses. Most organisms most of the time remain at close to ambient temperature, with all its regional and temporal variation. And some bacteria, for instance, tolerate truly infernal heat, reminding us that physics sets no real limit below the local boiling point of water. Still, laissez faire might make either chemical or physical trouble, or perhaps both.

Nearly all enzymatically catalyzed reactions depend severely on temperature. Rates typically double or triple for every 10°C rise in temperature, whether one looks at individual reactions or the overall metabolic rates of animals that do not regulate their temperatures. (To calculate proper temperature effects, the Arrhenius equation and Arrhenius constants should be used instead of this so-called Q_{10}.) On top of that, most enzymes, as proteins, denature with ever increasing rapidity as temperatures rise above around 40°C. For instance, one protein that denatures a marginally tolerable 4.4 percent per day at 40° cooks (to use the appropriate vernacular) at 46 percent per day at 46°.

As bad, perhaps, temperature-dependence varies from enzyme to enzyme, so sequences of reactions might demand something beyond simple mass-action effects to coordinate their operation. That may underlie the notably limited temperature range tolerated by many organisms. Most nonregulating inhabitants of habitats of nearly constant temperature cannot withstand body temperatures more than a few degrees above or below that normal range. They may be intolerant even when well above freezing and well below the range of severe protein denaturation. The extreme in sensitivity must be creatures such as ourselves that closely regulate body temperature. Such constancy typically brings a loss of ability to survive—even briefly and dysfunctionally—without it.

Most physical variables change less with temperature than do chemical reactions and metabolic rates. That same 10° rise in temperature (using 20° to 30° for the examples) decreases air density by about 3.4 percent and the surface tension of water (against air) about 2 percent. It decreases the thermal capacity of water (on a mole basis) a mere 0.04 percent but increases the diffusion coefficients of ordinary gases in air by about 6 percent. One notes as a benchmark that something proportional to the absolute temperature will increase by 3.4 percent for a 10° rise—as does the reciprocal of air density. For instance, Weis-Fogh (1961) showed that the tensile stiffness (Young's modulus) of the protein rubber of insect wing hinges, resilin, stiffens by just that 3.4 percent per 10-degree rise.

But a few physical quantities do vary more widely. The viscosity of water decreases by over 20 percent from 20° to 30°. An animal accustomed to pumping blood at 35°, say a reptile basking in the sun, must expend twice as much energy (or pump half as much blood) if it plunges into water at 5°—unless, as in so-called multiviscosity motor oils, its blood has a peculiarly low dependence of viscosity on temperature. Such an increase in blood viscosity at low temperatures might well compound its problems of the temperature-dependent decreases in basal metabolic rate and maximum metabolic capability. (Maximum capability may matter more since a drop in basal rate will decrease the need to move blood.)

Compounding the problem, the diffusion coefficients of solutes decrease as temperatures drop in parallel with the increase in solution viscosity. So for a given solute, the product of viscosity and diffusion coefficient will vary little with temperature. That recognition came as one of Einstein's great achievements during that *annus mirabilis*, a century ago, as he linked the viscosity (μ) in Stokes' law for small-scale flows with the diffusion coefficient (D) in the Sutherland-Einstein relation or Stokes-Einstein equation (Pais 1982):

$$D = \frac{RT}{N} \frac{1}{6\pi\mu r}, \qquad (4.1)$$

where R is the gas-law constant, T the absolute temperature, N Avogadro's number, and r the radius of the solute molecules. (William Sutherland obtained the same result in the same year, hence Pais's suggestion of a hyphenated name.) So both of the biological transport processes, diffusion and convection, will be seriously impeded in liquid systems by a drop in temperature. At least if flow slows in proportion to viscosity, then Péclet numbers (ratios of convective to diffusive mass transfer, as expanded upon in chapter 1) will not change, and system geometries ought still be appropriate. Put less encouragingly, tinkering with system geometry cannot easily compensate for temperature change.

Nor do viscosity and diffusion coefficients mark the extremes of temperature-dependence. Once again looking at a rise from 20° to 30°, the maximum concentration of water vapor in air (100 percent relative humidity) goes up from 17.3 to 30.4 grams per cubic meter—a 75.6 percent increase. Put another way, water vapor makes up a mass fraction of 1.44 percent of saturated air at 20° and 2.61 percent at 30°—an increase of 81.6 percent. No wonder a lot of water condenses on a cool body in a hot, humid environment.

HEAT-MOVING MODES

How might a creature move heat from one place to another, whether shifting heat from one inside location to another, absorbing heat from its surroundings, or dumping heat onto those surroundings? A rather large array of options turn out to be available.

(a) Radiation. All objects above absolute zero radiate energy. A net radiative transfer of heat from warmer objects to colder ones occurs even if the objects are in a vacuum.

(b) Conduction. Heat moves from warmer to colder parts of a material (or a contacting material) by direct transfer of the kinetic energy of its molecules.

(c) Convection. Heat moves from warmer to colder places by direct transfer of the warmer material itself. Ordinarily either cooler material or yet more material from elsewhere takes its place.

(d) Phase change. Vaporization takes energy, so it can absorb heat and leave a body cooler than otherwise. Fusion, likewise, takes energy, so melting a solid will cool either the rest of the solid or something else. Solid-to-gas change, sublimation, combines the two, absorbing even more energy.

(e) Ablation. The average temperature of an object of non-uniform temperature can be reduced by discarding some of its hotter-than-average portion, in effect exporting heat.

(f) Gas expansion cooling. A contained gas exerts some pressure on the walls of its container; if it pushes those walls outward, thus doing work, either its temperature will drop or it will absorb heat.

(g) Cooling by unstressing an elastomer. Stressing an elastomer (stretching rubber, for instance) warms it. Elastic recoil as it's released cools the elastomer.

(h) Changing the composition of a solution. Dissolving one substance in another—mixing two different liquids or dissolving a solute in a solvent—may either absorb or release heat.

Even without invoking ordinary chemical reactions or thermoelectric phenomena, we have at least eight modes of heat transfer, some of which can be divided further. All are reversible, and the last five can be used to move heat from something cool to something warm without offending thermodynamics. Physics assuredly affords an abundance of possibilities that we ought to examine for biological relevance.

RADIATIVE HEAT TRANSFER

The temperature of an object determines the peak wavelength at which it either absorbs or emits radiation. How it behaves at (or near) that wavelength depends on its emissivity and absorptivity; since these don't differ, a single measure serves, most often called the emissivity. (Were the two unequal, an isolated system might spontaneously move from temperature uniformity to non-uniformity, thermodynamically unlawful.) Not only peak wavelength but radiant intensity depends on temperature, the latter quite strongly; the Stefan-Boltzmann law gives the particulars:

$$q = \sigma S(\varepsilon_2 T_2^4 - \varepsilon_1 T_1^4), \tag{4.2}$$

where q is the rate of energy transfer, T_1 and T_2 the Kelvin temperatures of the objects involved in the radiative exchange, S the effective exposed area, and σ the Stefan-Boltzmann constant, 5.67×10^{-8} watts per the product of square meters times the fourth power of the Kelvin temperature. Note that it makes the necessary distinction between emissivities (the ε's) at incoming (= absorptivity) and outgoing wavelengths.

For that wavelength-dependence we turn to Wien's law (sometimes the Wien displacement law), asserting an inverse relationship between surface temperature, T, and peak emission wavelength, λ_{max}:

$$\lambda_{max} = \frac{0.0029}{T}. \qquad (4.3)$$

(The constant assumes temperatures in Kelvin and wavelengths in meters.) Thus peak emission of the sun, at about 5800 K, occurs at 500 nanometers, roughly in the middle of our visual spectrum. An organism at 30°C or 303 K will emit with a peak a little under 10 micrometers, far out in the infrared.

The solar peak at 5800 K, perceived by us as yellow, implies that both the photosynthetic machinery of plants and the visual systems of animals make good use of solar radiation. That may mislead slightly, an artifact of the way we ordinarily plot intensity against wavelength. The energy intensity represented by radiation varies inversely with wavelength, something we may mention parenthetically when cautioning against the ultraviolence one does to oneself by exposure to the ultraviolet. So a better picture emerges from a graph with a scale on its abscissa inversely proportional to wavelength and thus linear with respect to energy content. Wavelength inverts to frequency (f), with the speed of light, c, as conversion factor ($f=c/\lambda$), so frequency would work. In the usual practice, something called "wave number," the unadjusted reciprocal, $1/\lambda$, replaces frequency. Then equal areas under a line represent equal amounts of energy, wherever the areas might be located. Such a curve tolerates simple integration for energy, what matters when considering the heating effect of radiation.

Figure 4.1 gives that kind of a spectral graph for direct overhead solar illumination at sea level (data from Gates 1965, 1980), along with divisions into the usual regions (from Monteith and Unsworth 1990). Most of the energy we receive doesn't come in the visible at all! Fortunately, the ultraviolet makes up only a small component; infrared radiant energy actually exceeds visible radiant energy. (The various bands absorbed by water give the curve its jagged appearance in the infrared.) That infrared radiation can make trouble for terrestrial organisms.

Consider a leaf exposed to full sunlight. It must absorb solar energy to split water and fix carbon. Yet the photons of solar radiation at wavelengths longer than 700 nanometers are insufficiently energetic for that purpose. If absorbed, though, they would convert to heat. We rarely worry

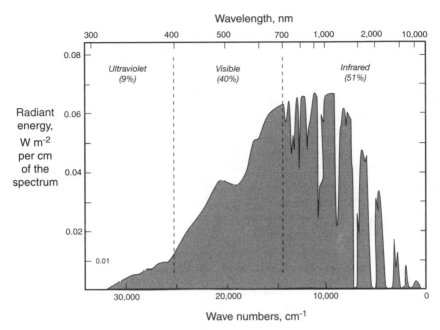

Figure 4.1. Spectral distribution of solar illumination of the earth's surface on a plot in which energy is uniformly proportional to area under the curve.

that leaves might get intolerably hot, but the possibility should not be dismissed. The 1000 watts per square meter of an overhead sun imposes no small thermal load—enough to heat a thin leaf by over 2° per second. By converting solar energy to a nonthermal form, photosynthesis might help, but its capture rate of 5 watts per square meter accounts for less than a percent of the load.

Leaves make a major dent in the problem by rejecting most of the infrared component of sunlight, and thus reflecting or transmitting rather than absorbing about half the overall input. Photograph foliage with infrared-sensitive monochromatic film and a red filter to stop most of the visible light, as in plate 4.1; leaves will appear white (on a positive) and, if the day is clear, the sky will be starkly black.

That white "color" should be regarded as something special. Ordinary pigments, fabrics, animal skin, and fur—all absorb infrared and thus look black. Among biological objects, are leaves unique in this respect? Unfortunately, radiative processes have drawn little attention from physiologists other than those concerned with terrestrial vascular plants. Bird eggs reflect most—sometimes over 90 percent—of the near-infrared. Bakken et al. (1978) showed the independence of an egg's visible color (commonly cryptic) from its infrared reflectivity as well as (in part) the basis of that reflectivity, the use of pigments other than the melanin typical of vertebrates. The shells and opercula of desert snails may also reflect most of the sun's direct infrared load (Schmidt-Nielsen et al. 1971; Yom-Tov 1971). And the spacing of the laminae in the cuticle of some iridescent red algae (from micrographs of Gerwick and Lang 1977) hints at infrared reflection. These organisms (*Iridaea* and others) can be exposed to both air and full sunlight at low tide, so the matter matters. Still, one suspects investigative inattention rather than biological rarity. Some practical technology might come from knowing a bit more—adding a truly white roof could reduce the internal temperature of a sun-lit house in a hot place, and a truly white-crowned hat might provide shade with less concomitant hot headedness.

The common lack of correlation between visible color and the overall absorption of solar radiant energy needs emphasis. Leaves and egg shells absorb little; fur of almost any color absorbs a lot. That may deprive us of an easy visual assessment, but it permits organisms to decouple color as seen by prey, predators, and conspecifics from effective radiant color.

In addition to receiving solar radiation, organisms exchange infrared radiation with their more immediate surroundings, with intensities and wavelengths set by the Stefan-Boltzmann (eq. 4.2) and Wien (eq. 4.3) relationships. Ultimately what matters is net transfer, something easy to forget when incoming greatly exceeds outgoing. One feels warmed on the side of the body that faces a surface above skin temperature, such as a stove, even when surrounded by air at a uniform temperature. Normally the temperatures and emissivities of organisms and their immediate surroundings are similar, so no great net heat transfer occurs. An exception is an open sky—a very large "object" at a low effective temperature. According to Nobel (2005), with clean air, the effective temperature of a clear night sky may be as low as 220 K (−53°C); with cloud cover that may rise to 280 K (+7°C). Thus something whose surface temperature is 30°C can radiate 3.6 times as much energy to a clear sky as it receives in return.

That asymmetry can be noticeable and significant. Since I became bald, I can feel whether a night sky is clear or cloudy without looking up, at least when no wind is blowing. If I stand still for a few seconds beneath a clear sky, I get a particular tingle in my scalp. Of more consequence is radiation from foliage. On clear, windless nights, condensation often forms frost on low plants even when the atmosphere well above the ground remains above the freezing point—the foliage radiates sufficient energy to the sky to drop its temperature below freezing and, by conduction and convection, that of the air in its immediate vicinity as well. The phenomenon can damage freeze-sensitive crops; prevention schemes include coverings, sufficient wetting to overwhelm the temperature drop from the radiant emission, or (at least formerly) burning smoky smudge-pots, not to heat the crop but to obscure the sky.

One small tree may take action to avoid exposure to that cold night sky. *Albizzia julibrissin*, sometimes called the silk tree, is native to China but well established as an ornamental in the U.S. southeast. Its doubly compound leaves with a few hundred leaflets give it a vaguely fernlike appearance. As shown in plate 4.2, the leaves seem to have three distinct postures. In the shade, both leaves and leaflets extend horizontally; a light shining down on a leaf is almost fully intercepted. In the sun, the rachis of the leaf remains horizontal, but leaflets shift to near-vertical so the leaf casts only a minimal shadow (Campbell and Garber 1980 describe the motor responsible). At night, the entire leaf bends down to near-vertical, with the individual leaflets folded against the rachis—the leaf then looks like the tail of a horse. I suspect that the orientation in direct sunlight reduces exposure to a point source of radiation while the complete folding at night reduces exposure to a distributed radiation sink.

Analogous postural control of solar irradiation is well known in terrestrial animals, particularly insects and lizards. Many of them can assume either postures that minimize solar input, for instance orienting vertically during the heat of the day, or postures that maximize solar input—or both, depending on the circumstances. Many insects absorb sunlight in preflight warm-ups that raise body temperatures well above ambient, taking advantage of their small size and consequently high surface-to-volume ratios. Wings often assist as shields against simultaneous convective cooling. (Heinrich 1996 gives an engaging account of the thermal devices of insects.) Lizards, larger, capitalize on their sit-and-wait mode of predation to engage in more leisurely thermal basking.

Some mammals control solar radiation as well. A ground squirrel (*Xerus inauris*) that inhabits hot, dry areas of southern Africa, for instance, turns its back to the sun when conditions get especially challenging. That puts it in a position so its tail can serve as a parasol to provide local shade. Bennet et al. (1984), who describe the behavior, calculate that the squirrel can thereby increase daytime foraging episodes from about 3 to 7 hours.

Organisms adjust emission as well as absorption. At the long wavelengths corresponding to their surface temperatures, desert plants have slightly higher emissivities than do plants from temperate regions, and those are slightly higher than the emissivities of plants from the rain forest (Arp and Phinney 1980). All values, though, are high, most above 0.95. At long wavelengths foliage in general, with emissivities of 0.96 to 0.98, emits more effectively than nonvegetated surfaces, typically about

0.91 (Kant and Badarinath 2002). What remains uncertain is whether that difference translates into a biologically significant additional heat loss.

Reradiation to the sky may underlie the peculiarly large and well-vascularized ears of many desert animals—jack rabbits (*Lepus* spp) in particular. As Schmidt-Nielsen (1964) points out, these animals are too small to cool by evaporating water, and most lack burrows as mid-day retreats. With air temperatures at or even above body temperature, their large ears look paradoxical. But by feeding in open shade, with hot ears exposed to a much colder sky (at an effective temperature of perhaps 13°C), such an animal could offload a lot of heat.

Conductive Heat Transfer

The formal rules for conduction of heat parallel those for diffusion. Fourier's law (eq. 4.5) renames the variables in Fick's law (eq. 4.4), using energy transferred per unit time (q) instead of mass transfer rate (m/t), temperature gradient ($T_1 - T_2$) in place of concentration gradient ($C_1 - C_2$), and thermal conductivity (k) rather than diffusion coefficient (D):

$$\frac{m}{t} = DS\left(\frac{C_1 - C_2}{x}\right) \tag{4.4}$$

$$q = kS\left(\frac{T_1 - T_2}{x}\right) \tag{4.5}$$

Here S is the area over which transfer takes place, with x the distance mass or heat has to move. In each process, a gradient—concentration or temperature—provides the impetus.

The only additional variable that matters is specific heat, usually given as c_p. For a given material, it establishes a proportionality between energy input relative to mass and change in temperature. Water has a fairly high specific heat, 4.18 kilojoules per kilogram-degree at ordinary temperatures, so a lot of heat raises its temperature only a little. For air c_p is lower, 1.01 kilojoules per kilogram-degree. For soils c_p is typically (but not inevitably!) about 1.0 to 1.5 kilojoules per kilogram-degree. Organisms, mostly water, rarely deviate much from its temperature-stabilizing high value.

For conduction through a slab of material, heat transfer varies inversely with thickness—as in equation (4.5). For gain or loss from a solid body, rates (for most geometries) run inverse with the square of linear dimensions—surface area, really. And just as some diffusive step underlies every case of transfer of mass by bulk flow (as noted in the first chapter), conduction plays some role in all convective processes.

(Advantage can sometimes be taken of that practical equivalence of diffusion and conduction. One can serve as proxy for the other, usually conductive heat transfer for diffusion because it's easier to measure temperature than chemical concentration—as done, for instance, by Hunter and Vogel 1986.)

Nowhere do nature and human technology diverge further than in their relative reliances on conduction. Humans have access to metals, materials of high conductivity; nonhuman nature uses no metallic materials, either within organisms or in their surroundings. Metallic and nonmetallic materials differ by orders of magnitude; table 4.1 gives a sampling of values. Between the low values of conductivity and its severe distance discount, conduction can play no great role in moving significant amounts of heat over appreciable distances within living systems. Again, consider a leaf in sunlight and nearly still air. The center of the leaf gets hotter than its margins because the latter make better thermal contact with the convective updraft induced by the hot leaf itself. Were the leaf made of metal, peak temperature would be lower—conduction would move heat down the temperature gradient from center to edges. But mainly air, water, and cellulose make up a leaf; and those nonmetallic materials can't move enough heat to affect that temperature gradient, unlike the metallic heat sinks with which we protect heat-intolerant semiconductors (Vogel 1984).

So one should not (as have several studies) use radiantly heated metallic models to study the thermal behavior of leaves. Such models will have lower center and average temperatures (Stokes et al. 2006). Perhaps more important, as a result of their lateral heat transfer they will approach the condition referred to in books on heat transfer as "constant temperature" rather than "constant heat flux." Most of their formulas assume that unbiological near-constancy of temperature since, unfortunately for the biologist, they're products of our metallic culture. Metal models are handy, but they have to be spot-heated in the middle rather than area-heated everywhere, and the thickness of metal must be chosen to give the center-to-edge temperature gradient of real leaves.

In a sense pure conduction represents a gold standard for minimal heat transfer. Thus, fur works by reducing convective air movement enough for overall transfer to approach the value for conduction in air. And heat exchangers (about which more in chapter 5) drive the heat transfer due to convective blood circulation down toward the value for conduction in a nonconvecting tissue such as a slab of meat.

Nonetheless, a few organisms do employ conductive heat transfer as more than a short-distance link between a flowing fluid and an adjacent

Table 4.1

Thermal conductivities of a variety of materials. Imagine heat transfer (watts per Kelvin) through a rectangular slab of material oriented normal to heat flow. The linear dimension, inverse meters, represents slab thickness (meters) over slab area (square meters).

Material	Conductivity $(W\,m^{-1}\,K^{-1})$
copper	385.
steel	46.1
glass	1.05
water	0.59
skin	0.50
muscle (meat)	0.46
adipose tissue (human fat)	0.21
wood (typical, dry)	0.20
soil (inverse with air fraction)	0.25 to 2.0
fur	0.024 to 0.063
air	0.024

surface. Our elderly house cat rests on dry straw in the garden on cool days; on hot days he shifts to some patch of bare soil or pavement out of direct sunlight. The behavior is common among those medium- and large-size domestic animals that have soft enough flesh and fur for effective contact with the substratum. More specific use of heat earthing has been documented in a desert rodent, the antelope ground squirrel (*Ammospermophilus leucurus*). For a diurnal desert animal it's especially small, so it heats up rapidly when foraging in the summer sun—0.2 to 0.8 degrees per minute. A squirrel deals with this heat load by tolerating brief bouts of hyperthermia (sometimes exceeding 43°C) and returning to its burrow as often as every ten minutes. In the burrow, it chills out rapidly by pressing itself against the walls, usually about 10°C cooler than its body (Chappell and Bartholomew 1981).

CONVECTIVE HEAT TRANSFER

Whether looking at thermal phenomena within or around our creatures, we can rarely ignore convective transfer. We saw that conduction poses few analytic problems, with reliable equations and only the peculiarities of biological geometries to complicate things. Radiative exchange may be less familiar, yet here too straightforward rules give guidance. By contrast, no such tidiness characterizes convection. While the textbooks for engineering courses (I particularly value Bejan 1993) provide reliable explanations, the equations they cite must be viewed warily. Most provide no more than rules of thumb, and many presume conditions quite different from what organisms encounter. Even the first figures of their three-significant-figure constants may diverge from our reality. A few of the complicating aspects of convection follow.

(a) Convective flows may be internal or external. We move lots of heat by pumping blood and other fluids through our various pipes and internal channels; flows of air and water around us transfer heat between ourselves and the environment. The basic phenomena may be the same, but the details and calculations depend strongly on whether the solid object surrounds the fluid or vice versa.

(b) Convective flows may be laminar or turbulent, with major differences for heat transfer. In most laminar flows (as in our capillaries) convection carries heat only in the overall direction of flow—conduction drives transfer normal to that direction. By contrast, the internal mixing of turbulent flow provides a major means of cross-flow transfer, and the thermal conductivity of the fluid loses most of its importance. For internal flows through circular pipes, the shift from laminar to turbulent occurs at a reasonably sharp value (2000 ± 1000) of a single variable that has already appeared in earlier chapters, the Reynolds number, Re:

$$Re = \frac{\rho l v}{\mu},\tag{4.6}$$

where ρ and μ are the fluid's density and viscosity, l the diameter of the pipe or width of the channel, and v its average flow speed. External flows may undergo similarly sharp transitions, but the location of the transitions depends a lot on texture and geometry—between Re's of about 20 and 200,000, with l now taken as a variously defined characteristic length of the object in the flow.

(c) Convective flows can be driven by density differences within the fluid—"free convection"—or they may be driven by some remote pumping of the fluid—"forced convection." Unlike the previous distinctions, regimes can be mixed. Another dimensionless number, the Grashof, Gr, provides an index of the intensity of free convection:

$$Gr = \frac{\rho g \beta (\Delta T) l^3}{\mu^2}.\tag{4.7}$$

The only new variable is β, the volumetric thermal expansion coefficient. All gases have about the same value of β. Since their volumes vary directly with the absolute temperature, $\beta = 1/T$; at 20°C, $\beta = 3.4 \times 10^{-3}$ inverse degrees. The value for liquid water at ordinary temperatures is about 0.3×10^{-3} inverse degrees—about ten times less. Free convection mainly matters for external flows. It can be either laminar or turbulent, with a transition from former to latter at a Grashof number of about 10^9.

For big objects and objects far from ambient temperature, free convection looks important—but what about the role of ambient flow speed? A more complete picture emerges when Reynolds number provides a comparison. So a better, if still rough-and-ready, criterion can be had from another dimensionless index, the ratio of the Grashof number to the square of the Reynolds number. In effect, this looks at the ratio of buoyant force to inertial force; viscous force, affecting both components, cancels out. Thus

$$\frac{Gr}{Re^2} = \frac{g\beta(\Delta T)l}{\rho v^2}. \tag{4.8}$$

Some sources give the following rules of thumb. For ratios below about 0.1, forced convection predominates and free convection can be ignored. For ratios above about 16, free convection predominates and the effects of any wind or water current can be ignored. Higher thermal expansion coefficients, larger differences in temperature between organism and surroundings, and larger size raise the value and favor free convection; denser fluids and more rapid flows favor forced convection. All of which seems intuitively reasonable.

By this criterion, the biologist can't ignore mixed regimes. Once again, consider a sun-lit broad leaf on a tree. A leaf 10 centimeters across will face a mixed regime at wind speeds between about 0.04 and 0.5 meters per second. The lower figure is less than ambient wind ever gets for more than a few seconds in full sun. If nothing else, differential heating of ground and other foliage will generate that much convection. The higher figure, about our perceptual threshold for air movement, will nearly cool a leaf to air temperature—stronger winds make little further difference, and overheating ceases to be hazardous.

For that leaf, then, the thermally significant regime is a mixed one, the regime least amenable to anything other than direct measurements. Some years ago, a local engineering graduate student, Alexander Lim (1969), compared published formulas for mixed free and forced convection with measurements under conditions a leaf might encounter. He found shockingly great deviations, with discrepancies running around 50 percent—in both directions. And he controlled variables that in nature would confound things even further. Among the complications, free convection carries air vertically, while forced convection need not be horizontal since it includes not just ordinary wind but the upward free convection of adjacent leaves.

The main generalization one might make is that free convection will be insignificant for very small systems and a major consideration only for quite large ones. As Monteith and Unsworth (1990) point out, a cow might lose heat by free convection when the wind drops below about 1 meter per second. Similarly, at night a camel need not wait for a gentle breeze to dump heat convectively that it had acquired during the previous day. Judging from photographs of thermal updrafts around standing humans (using a technique which visualizes differences in air density), our large size permits some self-induced free convection. Still, even barely perceptible air movements help us avoid overheating when we work hard under hot and humid conditions. At yet larger scales free convection becomes yet more important; together with spatially irregular heating of the ground it produces the ascending thermal tori in which birds such as hawks and vultures soar.

CONDUCTION VERSUS CONVECTION

Since biological systems are made of low conductivity materials, pure conduction with zero convection amounts to a gold standard for minimal internal heat transfer. A warm human increases convective transfer by vasodilation of capillaries in the skin and the associated larger blood vessels—body temperature becomes less spatially variable. When cold, one reduces blood flow to the extremities, setting up internal temperature gradients closer to those of conducting systems. But we humans remain convection-dominated, reflecting both our high aerobic capacity and warm-climate ancestry.

How might one determine the relative importance of conduction and convection in an intact, living animal? Measuring blood flow will not give reliable results since heat exchangers (about which more in the next chapter) can decouple heat flow from mass flow. A simple scaling argument suggests at least one possible approach—it adopts the rationale for circulatory systems of the Nobel laureate physiologist August Krogh (1941), merely substituting heat for oxygen.

If on the one hand heat content depends on volume ($\propto l^3$) and heat loss depends on surface area ($\propto l^2$), then the rate of heat loss relative to volume will vary inversely with the first power of a typical linear dimension ($\propto l^{-1}$). That should happen where heat moves much more readily within the organism than to or from the organism. If on the other hand heat loss depends on the distance between core and periphery ($\propto l^1$), then the rate

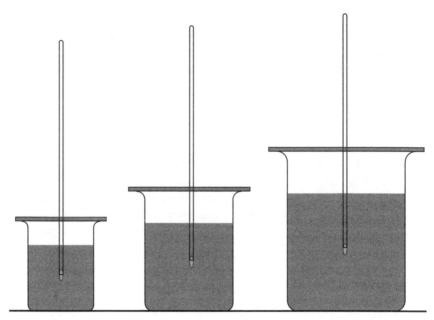

Figure 4.2. The arrangement of beakers and thermometers used to obtain the scaling data of figure 4.3.

of heat loss relative to volume will vary inversely with the square of linear dimensions ($\propto l^{-2}$). That will happen when heat transfer to and from the organism presents less of a barrier than transfer within the organism. Muscle, fat, and other biological materials have low thermal conductivities (table 4.1 again), while circulating liquids make fine heat movers. So the first situation (loss $\propto l^{-1}$) will characterize convection-dominated cases, the second (loss $\propto l^{-2}$) conduction-dominated cases.

At least at this crude level of judgment one needs only scaling data—another use of such data. Measurements of core body temperatures as equilibrated animals are heated or cooled will suffice, at least for ectothermic animals—temperature tracks heat loss per unit volume if heat capacity remains constant. Can such an easy model apply, or do confounding factors overwhelm it?

As a quick test, I created two sets of heat-transferring systems, one predominantly convective and the other exclusively conductive. Each set consisted of six ordinary laboratory beakers, of nominal capacities from 50 to 1000 milliliters, with each beaker filled to a depth equal to its internal diameter. A thermometer supported by a piece of corrugated paper-

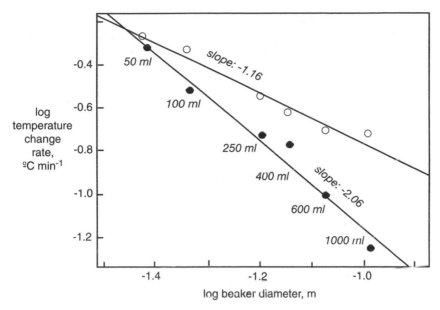

Figure 4.3. Log cooling rate versus log beaker diameter for water-filled (slope = −1.16) and water-plus-agar-filled beakers (slope = −2.06). The slopes represent the exponent *a* in the expression [cooling rate ∝ diameter^a].

board was extended down to the center of each beaker, as in figure 4.2. One set contained pure water while the other was filled with water plus 1 percent agar. That small amount of agar suffices to immobilize the water, preventing the free convection of self-stirring without significant effect on its specific heat. I equilibrated the twelve beakers overnight in an incubator at 49°C, moved them at time zero to a room at 25°C, and recorded their temperatures every 5 minutes. Free convection stirred the water-filled beakers enough to make deliberate stirring unnecessary, and room air movement sufficed to minimize external resistance. Figure 4.3 shows the results, with log-log slopes satisfyingly close to the predicted values. An analogous exercise in which the beakers warmed after equilibration at 7°C gave much the same result. Either way, immobilizing the water gave greater reductions in the rate of temperature change in larger systems.

So this simplest of scaling rules can place systems on a spectrum from virtually pure conduction to predominant convection. Turning to the living world, we might look at some old data for cooling lizards. For a variety of cooling varanids, Bartholomew and Tucker (1964) found a scaling exponent of −1.156 (tripling their mass-based number to change the abscissa to log of length). That's just a bit greater than what we would expect for convection-dominated systems. By contrast, Bartholomew and Lasiewski (1965) reported an exponent of −1.881 for Galapagos marine iguanas suddenly immersed in cold water, just short of what we anticipate for conduction-dominated systems. During dives, heart rates slow, but no more so than for the varanids. Somehow they must reroute their blood so it carries little heat peripherally. (Whether in air or water, the iguanas reheat much more rapidly.)

Cooling slowly makes adaptive sense for reptiles that bask on warm, sunny, shoreline rocks and then feed into cold water. Not that these iguanas do anything unprecedented. Immersed reptiles quite commonly heat faster than they cool, with the ratio increasing with body size, as noted by Turner (1987) and consistent with our scaling exponents and rationalization. Charles Darwin gives a fine descriptions of iguana and its behavior in *The Voyage of HMS Beagle* (1845) ("a hideous looking creature") as well as in his diaries (unoriginally, "imps of darkness")*. To my eyes they look less malevolent; the reader can judge from plate 4.3. Still, de gustibus non est disputandum.

One caveat (to stick with Latin). For systems surrounded by moving air—uninsulated systems with either free or forced convection—the external resistance to heat transfer can drop well below the internal resistance. Thus, reducing internal resistance by augmenting convection within the system may not decrease the exponent for loss relative to size as much as expected. Cooling infertile eggs, within which convection is negligible, provides an example, if a somewhat unnatural one. Turner (1988) obtained a heat loss exponent of −1.2 when in still air, which suggests internal convection. In fact it just reflects free convection in the external air. Putting them in water, which minimizes external convection, lowers the value to −1.8, nearly the −2.0 of our model.

HEAT TRANSFER BY EVAPORATION AND CONDENSATION

Vaporization of a liquid (or sublimation of a solid) provides a particularly effective heat-transfer mechanism, especially if, as does water, the liquid has a high heat of vaporization. Indeed, the value most often found in nonbiological sources, 2.26 megajoules per kilogram, presumes boiling at 100°C and understates the case; at a more biologically reasonable 25°C, water's heat of vaporization is 2.44 megajoules per kilogram, about 8 percent higher.

---

*In Darwin's day, "imp" carried a sense of the diabolical rather than, at present, merely of the mischievous.

Several conditions, though, limit its use by organisms. The atmosphere into which water vaporizes must not be water-saturated, at least at the temperature of the evaporating surface (which, as our skin normally is, may be above ambient temperature). Evaporation itself will reduce the temperature of the evaporating surface (again, as with our skin). And a copious supply of water must be available. In its typical warm, dry habitat a succulent plant, some lore to the contrary, cannot store enough water for significant evaporative cooling. Cooling evaporatively as we do, hard-working humans must consume water at a great rate. At a metabolic rate of 400 watts (a minimal estimate for a laborer working at 100 watts output), dissipating metabolic heat by evaporation, our main mechanism, would take 0.6 liters per hour, or almost 5 liters for an eight-hour working day.

Few small animals can rely on evaporative cooling as a principal mode during sustained activity—it demands too great a volume of water for the surface area exposed to a hot environment or for their metabolic rates (which scale nearly with surface area). Fortunately their higher surface-to-volume ratios improve the efficacy of convection. Even without deliberate evaporative thermal control, they seem more often concerned with water conservation, with devices that reduce respiratory water loss and so forth.

Among animals that cool evaporatively, two routes play major roles. Each has its points. Evaporation from skin (predominant in humans, cattle, large antelopes, and camels) takes advantage of their large areas of skin. The concomitant vasodilation improves convective loss as well. On the debit side, cutaneous evaporation inevitably causes a loss of salt, which then becomes a particularly valuable commodity for most herbivores that are active in hot climates. In addition, its requirement for exposed external surface conflicts with the presence of fur or plumage that might reduce heat loss under other circumstances.

Respiratory evaporation entails no salt loss, but it requires pumping air across internal surfaces, which costs energy and produces yet more heat. Furthermore, the CO_2 loss in excess breathing drives up the pH of the blood. Some animals such as dogs, goats, rabbits, and birds use respiratory evaporation, beyond that associated with normal gas exchange. They reduce both pumping cost and CO_2 loss by panting—shallow breaths repeated at rates matching the natural elastic time constants of their musculoskeletal systems (Crawford 1962; Crawford and Kampe 1971).

Some mammals (rats and many marsupials) cool evaporatively by licking their fur and allowing the saliva to evaporate; that mode, though, is not used during sustained activity. Some large birds (vultures, storks, and others) squirt liquid excrement on their legs when their surroundings get hot (Hatch 1970). They thus augment evaporative cooling the same way as modern air conditioners that drip condensate from inside on their external exchangers. A few insects (some cicadas, sphingid moths, and bees in particular) with ample access to water (nectar and sap feeders) benefit from evaporative cooling for dumping the heat produced by flight muscles despite their low surface-to-volume ratios (Hadley 1994; Heinrich 1996).

What about leaves? Again, many do get well above ambient temperature, pushing what look like lethal limits. Plants with broad leaves, the ones likely to run into thermal trouble, evaporate water ("transpire") at remarkable rates. Leaf temperatures calculated (from admittedly crude formulas) by Gates (1980) point up the thermal consequences of that evaporation. He assumed a wind of 0.1 meters per second (as noted earlier, about as still as daytime air gets), solar illumination of 1000 watts per square meter (again, an unobstructed overhead sun), an air temperature of 30°C, a relative humidity of 50 percent, and a leaf width of 5 centimeters. If reradiation were the only way the leaf dissipated that load, he figured that it would equilibrate (recall equation 4.2) at a temperature of about 90°C. Allowing convection as well drops that to a still stressful 55°C. A typical level of evaporation cools the leaf to about 45°C—hot but not impossibly so for a worst-case scenario. Evaporation cools leaves; it could not do otherwise. Broad leaves typically dissipate about as much energy evaporatively as they do convectively, but with great variations in the ratio both between kinds of leaves and even between locations on a given tree.

The thermal role of this transpirative water loss remains less clear than its thermal consequences. Plant physiologists (see, for instance, Nobel 2005) generally regard the loss as an inevitable by-product of the acquisition of carbon dioxide. They note that a leaf with openings (stomata) that admit inward diffusion of CO_2 will permit outward diffusion of water. CO_2 makes only about 0.03 percent of the atmosphere, and the diffusion coefficient of CO_2 is well below that of water. So a lot of H_2O must vaporize for even a modest input of that crucial carbon upon which plants depend. A representative value for water use efficiency (Nobel 2005) is about 6 grams of CO_2 gained per kilogram of H_2O lost. Functioning leaves have to lose water, whatever the thermal consequences. Indeed, transpiration sometimes depresses leaf temperatures 10°C or more below ambient even where temperature control should not matter. The situation resembles evaporative heat loss from our breathing, something of minor use (since we don't pant) for an excessively warm human yet a distinct liability for one stressed by cold.

But that view cannot be wholeheartedly embraced. Water use efficiencies vary too widely. The extreme values come from measurements on those species (6 or 7 percent of all vascular plants) that only open their stomata at night, when temperatures are lower and relative humidities higher. They fix CO_2 as organic acids; decarboxylation the next day then provides the input for photosynthesis. That trick can push up water use efficiency by an order of magnitude. So the adaptive significance of evaporative water loss from leaves remains uncertain. The question has drawn little attention—plant physiologists have worried less than have animal physiologists about primary—adaptive—versus secondary functions of multifunctional processes.

If evaporation cools, then condensation heats. Under at least one condition organisms may use condensation as a significant heat source. On cold, clear, calm nights, radiative cooling, as noted earlier, often drops leaf temperatures below both the local air temperature and the local dew point—the term "dew point" comes from the resulting condensation. Sometimes water vapor condenses as frost; where that happens the heat of sublimation, greater by 13 percent at 0°C than the heat of vaporization, becomes the relevant factor. Condensation clearly provides a major water source for some low desert plants. As new dew or fresh frost it must offset some of that radiative cooling; again, the practical significance is uncertain. Frost per se causes little trouble—what damages plants is the internal ice formation commonly signaled by its appearance.

A wide variety of arthropods—most notably some beetles of the Namib, an unusually humid desert (Parker and Lawrence 2001)—have been shown capable of condensing water from the atmosphere. In none does it seem to be such a simple physical process—the required temperature differences just do not occur, nor would they be likely in animals as small as ticks, fleas, and mites. Nor is a vapor-saturated atmosphere necessary; the minimum humidity can be as low as 50 percent. In none of these animals does condensation appear to confer any specific thermal benefit. Rather, obtaining liquid water is the payoff (Hadley 1994). We'll get back to that particular gambit in chapter 9.

A recent report implies a thermal role for still another form of phase change, one whose novelty may only reflect oversight earlier. According to Dunkin et al. (2005), a large fraction of dolphin blubber consists of fatty acids with melting points just below body temperature. The apparent thermal conductivity of the blubber of both young dolphins and pregnant females is well below that of human fat (as in table 4.1). Heat flux measurements suggest heat absorption by phase change as the mechanism, heat that would otherwise be lost to the ocean.

OTHER MODES—KNOWN AND UNKNOWN

So far, we've only looked at half the heat-transfer modes mentioned at the start—radiation, conduction, convection, and phase change. Some of

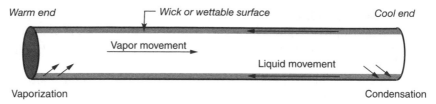

Figure 4.4. The operation of a heat pipe, with heat flow from left to right.

the others either can be dismissed outright or their insignificance easily argued. Early spacecraft used *ablative cooling* when reentering the atmosphere. Animals, as noted, do void saliva and excrement, but the subsequent evaporation of the liquid from deliberately wetted skin or fur does far more to get rid of body heat than does ablation itself.

You can use *gas expansion cooling* to make excessively hot food or drink palatable by pursing lips and exhaling air that has been compressed by your thoracic muscles—air temperature can be dropped into the mid 20°s according to a quick measurement on a cooperative colleague. But the muscle-powered compression-expansion sequence heats you more than it cools the food, so it's no good as a whole-body scheme. Significant heat transfer by *stressing and unstressing elastomers* seems unlikely, even if the imperfect resilience of biological materials might provide a small supplement to muscular heat generation under conditions such as the preflight warm-up of insects or our shivering. Similarly, transferring significant amounts of heat by *dissolving or extracting solutes* is unlikely, even though organisms commonly manipulate the composition of solutions.

What ought not be casually dismissed are novel combinations of the various heat-transfer mechanisms. As an example of an unknown but biologically plausible scheme, consider a so-called heat pipe (figure 4.4), a device that combines phase change and convection. A liquid vaporizes at the warm end, absorbing heat. Vaporization produces a pressure difference that drives gas toward the cool end. There it condenses, releasing heat. Liquid then returns to the warm end by capillarity through some wicking material lining the pipe. A few uncommon bits of human technology use heat pipes since they can achieve effective conductivities even greater than that of copper bars of the same dimensions, but they have never become household items.

By contrast, heat pipes should be highly advantageous in nature because of the low thermal conductivities of the materials of organisms. Having only water as a working fluid, though, imposes a serious limitation. Admittedly water has a nicely high heat capacity. And the concentration of vapor at saturation is strongly temperature-dependent; recall the 81.6 percent increase in mass between 20° and 30°C that was mentioned earlier. But pressure-driven bulk flow from warm to cool end cannot drive vapor movement as it does in systems where nothing dilutes the functional substance. Rough calculations suggest that diffusion, the obvious alternative, will not move enough water vapor over distances greater than about a millimeter to do much good. So such a system requires local stirring of the gas phase—cross-flow thermal gradients, continuous flexing of the pipe, or something else.

Where in organisms might we find heat pipes? Air-filled passages with hydrophilic inner surfaces are not rare. I wonder about the insides (the spongy mesophyll) of small, succulent leaves. Several colleagues, Catherine Loudon and Thomas Daniel, suggest that insects might use the mechanism to move heat from flight muscles through their tracheal systems.

So the case with which we measure temperature may too easily obscure the complexity of thermal phenomena in general and of the thermal behavior of organisms in particular. Designing proper experiments challenges our ingenuity. For instance, putting an organism in a temperature-controlled chamber may not come close to mimicking the thermal character of a habitat of the same temperature. The walls will not behave like open sky, and the heat source will be unlikely to resemble the sun. The air movement needed to ensure constant temperature will probably be unrealistically high—for instance for studying thermally stressed leaves. Or it may be unrealistically low—for, say, looking at the insulation fur provides for a mammal in the open. Beyond these difficulties lie the problems associated with continuous variation in environmental temperatures, insolation levels, wind speeds, and so forth in nature.

Put in these terms, the obstacles appear daunting. I prefer to view the situation in a different light. Physical complexity instigates biological diversity, not just in phylogenetic terms, but as diversity of clever designs and devices awaiting elucidation. And identifying what nature does begins by recognizing the physical possibilities.

CHAPTER 5

Maintaining Temperature

THE HAZARDS OF FEVERS AND CHILLS

Little else around us varies as much as the thermal loads that we terrestrial organisms face. Too often we're too hot, too cold, too well illuminated by sunlight, too exposed to an open sky, or in too great contact with hot or cold solid or liquid substrata. Beyond that, both soil and water temperatures may be far from constant. Thermal loads vary in time scale as well as in magnitude. Air temperatures and radiative regimes change over every time scale relevant to their operation, from seconds to years, at the least. Variation may be as assured as (and because of) night following day or it may be predictable only in a general statistical sense. Terrestrial life—and sometimes even aquatic life—is a hotbed of thermal challenges.

The last chapter argued that variable internal temperature could impose serious constraints on biological design. It looked first at the way temperature, both extremes and fluctuations, might affect the operation of organisms. It then turned to the various physical agencies that could move heat to, from, and within organisms. Here I'll take a complementary look at these same issues, exploring the ways in which organisms mitigate those fluctuations, focusing in particular on how creatures can avoid moving heat.

Ideally, holding one's internals at a different temperature from that outside should cost no energy—cost reflects imperfect thermal isolation. We might venture a sweeping generalization and assert that adaptations for maintaining appropriate temperatures in a world of extremes and fluctuations have a particular common character. All (or, to be on the safe side, almost all) work so as to minimize the metabolic cost of temperature control. While energy economy may not be the transcendent issue that many of us once presumed, its importance can't be denied.

And we might reassert another generalization, a bit less sweeping than the preceding one, from the previous chapter. Conduction, whether through air, water, or tissue, most often establishes a baseline; pure conduction usually will represent the ideal. For transfer within an organism, the central challenge comes down to reducing the convective heat transfer accompanying flow in blood vessels and air passageways to a level at which conduction predominates. If that can be done, then avoiding excessive temperature fluctuations with minimum energy expenditure can take advantage of the conveniently low thermal conductivities of life's two main media, air and water. Or, much the same as the latter, flesh and bone.

Thus air and water set the standards. All gases have low thermal conductivities; air's value, 0.024 watts per meter-degree, is ordinary for a gas or gas mixture—argon may be 32 percent and CO_2 36 percent lower, but hydrogen is 7 times higher. Liquid water, at 0.59 watts per meter-degree, is quite as ordinary, here by comparison with other nonmetallic liquids as well as with solids—40 percent lower than glass and 46 percent lower than limestone but about three times higher than pure fat, isolated whale or seal blubber (see Dunkin et al. 2005), and common plastics such as the acrylics. Except, perhaps, for switching from watery muscle to minimally hydrated fat, we organisms have little scope for reduction of thermal conductivity.

(Instead of thermal *conductivity*, animal physiologists often use thermal *conductance*, the combined rates of conductive and convective transfer per unit surface and per degree. With units of watts per square meter-degree rather than watts per meter-degree, it ignores the thickness of any insulating layer. That makes good sense when looking at experimental data from irregularly shaped and variably coated animals. By contrast, data for conductivity usually come from in vitro measurements on pelts and tissue samples. In short, invariant thermal conductivity implies that thermal conductance varies inversely with layer thickness.)

CIRCUMVENTING CONVECTION

Unless prevented by some specific device, air ordinarily moves, and it moves at speeds that matter. Speeds far less than our perceptual threshold of about 0.5 meters per second still have thermal consequences. An oak leaf in the sun with an axial temperature of 41°C at 0 to 0.01 meters per second will reach only 37° at 0.1 meters per second (Vogel 1968). So until shown (or calculated) otherwise, still air in the meteorological sense should be presumed nonexistent in the thermal world of organisms. If nothing else, any organism whose surface temperature differs from the surrounding air will experience self-induced free convection. Additionally, macroscopic organisms move fluids internally since some form of bulk transport is a practical prerequisite for getting much above cellular size (chapter 1). Such transport systems will move heat as well as material, heat transfer with either positive or negative consequences.

One way to reduce internal convective heat transfer simply consists of reducing blood flow to the periphery and extremities by vasoconstriction.

Small adjustments in relative vessel diameters can substantially reroute blood. We certainly do just that when inactive and exposed to cold, allowing our skin and appendages to stay at temperatures well below those of our brains and viscera. In the cold, skin temperature is normally 10° below that of the body's core. In one old experiment (DuBois 1939) nude males were asked to rest quietly in what was described as still air. Exposure to an ambient 22.5°, perceived under the circumstances as quite chilly, dropped core temperature by about 0.5°. By contrast, average skin temperature dropped 7°—hands somewhat less, feet by as much as 10°. In more extreme cold exposure we begin to defend core temperature by increased metabolic activity, noticeable as the minimally coordinated muscular contractions of shivering, rather than by further reduction in peripheral circulation. These responses appear fairly general among warmblooded vertebrates, not just unfurry and unfatty ones such as ourselves.

In practice, vasoconstriction combines two physical agencies. It reduces convection by creating a peripheral region in which flow is minimal. And it lowers conduction because lengthening the distance between central and surface temperature in, say, an appendage reduces the steepness of the temperature gradient. Experimental studies rarely tease apart the mix, but one presumes that it varies case to case and place to place.

Adding insulation works much the same way as vasoconstriction, again through a pair of physical agencies. It has two biological manifestations—internal insulation using peripheral layers of fat and external insulation of fur and feathers.

Fat, as noted earlier, has an agreeably low thermal conductivity, about three times lower than water or meat. In addition, few tissues approach the low metabolic activity of subcutaneous fat—the reason metabolic rates are often referred to as "lean body mass" for comparisons among different animals. Thus, addition of subcutaneous fat reduces peripheral circulation as well as adding insulation. Subcutaneous fat layers can be remarkably thick, approaching 50 percent of total body volume in aquatic mammals that swim in cold waters. With this blubber, a seal can simultaneously have a skin temperature about a degree above 0° and a core temperature in the mid-30°s (Irving and Hart 1957). The significant insulating effect of subcutaneous fat in humans underlies the common observation that females, with thicker layers, tolerate full-body exposure to cold water better than do males, whether they are Korean pearl divers or (as I've observed) marine biologists.

Fur and feathers permit effective conductivities to approach the value of pure air by limiting both free and forced convection. In no case does their own conductivity, that of the protein keratin, take on particular importance. Again we lack good data on how much of their effectiveness represents restriction of flow (usually air, in this case) and how much comes from reduction in the thermal gradient over which conduction occurs. Another uncertainty concerns the effects of ambient wind. The design of a fur coat of greatest effectiveness for its cost and thickness must depend on the relative importance of the free convection of the warm animal itself and the forced convection of environmental air movement. A fur coat has a dynamic component as well. Piloerection permits adjustment of its thickness and thus its thermal effectiveness—although our own attempts, noticed as so-called gooseflesh or goose bumps (reminiscent of plucked poultry) don't accomplish much.

In practical terms (sweeping such complications aside) a single number provides a simple measure of the effectiveness of the fur coat of a mammal. That's practical because of the near-uniformity of mammalian core temperatures and our consistent preference for insulation over metabolic increase. One needs only the temperature below which insulation alone won't permit a mammal or bird to maintain normal body temperature at basal metabolic rate. Put another way, that's the temperature below which extra food must be expended simply to stay warm, the "lower critical temperature." Naked human males (females, again, with more subcutaneous fat, do a bit better despite their smaller sizes) have to turn up the fire at about 27°, not at all impressive—presumably we're still warm-country pursuit predators, better adapted for dissipation than conservation of heat. By mammalian standards, we do poorly—although sloths do still worse, with critical temperatures around 29°. Even small mammals, with fur length limited by such things as their need for feet that reach down to the ground, do better. For instance, weasels manage 17° and ground squirrels 8°. Large mammals, especially arctic ones, tolerate cold with remarkable economy—lower critical temperatures commonly run between 0° and −40° C. These data come from Scholander et al. (1950).

OFFSETTING CONVECTION WITH COUNTERCURRENT EXCHANGERS

Counterintuitively, a convective link between hot and cold locations need not transfer heat. The agency can be turned against itself—if it can carry heat one way, it should carry it the other quite as well. In the present context, a warm animal in a cold place, the trick consists of transferring heat from blood flowing peripherally, not to the environment, but to blood flowing axially. The engineering literature refers to the operative device as a "counterflow" exchanger, physiologists prefer the word "countercurrent" (often spelled "counter-current"). The key element is a region, typically near the base of an appendage, in which arteries and veins lie in sufficiently intimate juxtaposition for that heat transfer. If blood were to travel in the same direction in both arteries and veins, the best that could be achieved would be an output that averaged hot and cold inputs. But a counterflow arrangement, as in figure 5.1a, runs into no fundamental limit on transfer; practical limits depend on the intimacy of the vessels, the flow rates, the conductivity of blood and vessel walls, and the outer insulation of the exchanger. Exchange is not limited to heat—diffusion, again, follows the same rules as conduction—and countercurrent exchangers also conserve such substances as dissolved oxygen and water.

Figure 5.1b shows a device with which students in a course I once taught explored the operation of such exchangers. I asked them to compare two exchangers, a countercurrent one in which flows ran in opposite directions (as in the figure) and one in which (with a pair of connections reversed) flows run concurrently. We quantified their deficiencies as the difference in temperature between input (T_{in}) and output (T_{out}) divided by the overall temperature difference between hot and cold ends (ΔT); subtraction from unity expressed data as exchange efficiency, e_e:

$$e_e = 1 - \frac{T_{out} - T_{in}}{\Delta T}. \tag{5.1}$$

For both exchangers, plots of efficiency against flow speed showed, as expected, that faster flow reduced the effectiveness of exchange while slower flow gave better performance. But even at very low flow speeds, the concurrent exchanger never quite reached 50 percent efficiency, while even this crude countercurrent device easily exceeded 90 percent.

The recognition of countercurrent exchangers in organisms has a curious and instructive history. The closeness of large arteries and veins in our appendages had been recognized for over three centuries before the father of physiology, Claude Bernard, in 1876, suggested that the combination might work as a heat exchanger. And early comparative anatomists had described many cases of local branching arrays of vessels that formed networks of intermingled arteries and veins. They called such a structure a "rete mirabile" (plural "retia"), literally a "wonderful net," or a "red gland" for the color imparted by all that blood. Retia were recognized by, among others, Francesco Redi (1626–1697), still remembered for his experimental evidence against the spontaneous generation of maggots from meat. J. S. Haldane (father of the more flamboyant J.B.S. Haldane) in his classic book, *Respiration* (1922), had the right idea about the fish swimbladder rete (an exchanger of dissolved gas, not heat). He drew an analogy with "a regenerating furnace, where the heat carried away in the waste gases is utilized to heat the incoming air."

Somehow the common function of these retia escaped notice. Why? Traditional anatomists did not think in either functional or nonbiological

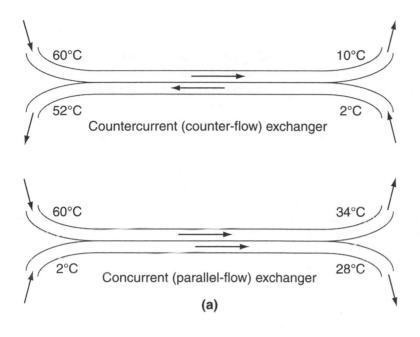

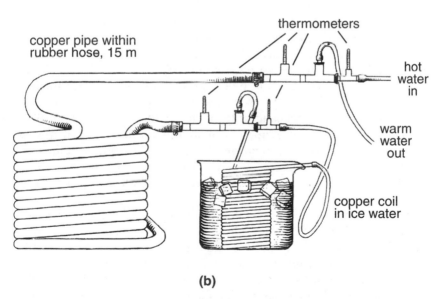

(b)

Figure 5.1. (a) Two heat exchangers, one in which fluid in the two channels flows in opposite directions and another in which it moves in the same direction. The temperatures represent typical results obtained by students using the device in (b). (b) A device that can be used as either a countercurrent (as here) or a concurrent exchanger. It consists of axial and annular channels and is made of flexible copper household tubing about 1.5 centimeters in diameter, rubber automotive heater hose about 3 centimeters in diameter, copper plumbing fittings, and laboratory stoppers, thermometers, and tubing.

terms. Physiologists, with only rare exceptions, focused on particular functions and particular animals, mainly humans, who happen to lack blatant examples of such exchangers. Take your pick of explanations. But once someone drew sufficient attention to the basic function of a rete, practically every known instance was quickly reexamined and assigned a functional role. Variants appeared, as did exchangers of less definitive anatomical character and less efficient operation. For instance, two veins (venae comitantes) surround the brachial artery of our upper arms, forming the exchanger noted by Bernard. The trio develops a lengthwise thermal gradient, though, of only about 0.3 degrees per centimeter, and we conserve more heat by shunting blood away from superficial vessels (Bazett et al. 1948) than by their action.

Retia, then, have long been known. The person who first drew general attention (one wonders whether the word "discovered" applies) to how they worked as countercurrent exchangers that could conserve either heat or diffusible molecules was an especially creative physiologist, Per Scholander (1905–1980) in the 1950s. He credited Haldane, who credited Redi and others. It was his 1957 article in *Scientific American* that seems to have started that transition from obscurity to fashion. The first formally described function was not heat exchange but transferring dissolved gas in the vessels supplying the swimbladder of deep-sea fish (harkening back to Haldane). The device allowed them to secrete and maintain gas in the bladder, gas that pressures of up to several hundred atmospheres should redissolve and return to blood, gills, and then ocean (Scholander and Van Dam 1954). The flukes of small whales provided the first definitive examples of heat exchangers (Scholander and Schevill 1955). Blubber, noted earlier, provides superb insulation, but thickly coated appendages would be ineffective propulsors. Exchangers allow these animals to supply effectively cold-blooded fins with blood from an otherwise warm-blooded body without a futile investment of metabolic energy in heating the global ocean.

Highly effective countercurrent heat exchangers have now been described in the bases of the appendages of sloths, anteaters, and some lemurs (Scholander and Krog 1957), the legs of wading birds (Scholander 1955; Kilgore and Schmidt-Nielsen 1975), the tails of muskrats (Irving and Krog 1955; Fish 1979), beavers, and manatees (Rommel and Caplan 2003), the legs of leatherback turtles (Greer et al. 1973), and the testicular blood supplies of marsupials, sheep (Barnett et al. 1958), bulls (Glad Sorensen et al. 1991), and dolphins (Rommel et al. 1992). They isolate the warm, dark, lateral muscles of large, fast-swimming tuna and mako sharks from the colder water passing along the body and across the gills (Carey and Teal 1966, 1969; Dewar et al. 1994). Gazelles, sheep, and some other ungulates keep their brains from getting as hot as the rest of their heat-stressed bodies with a carotid rete, where ascending arterial blood is cooled by venous blood coming from evaporatively cooled nasal passages (Baker and Hayward 1968). Honeybees and some other Hymenoptera isolate their abdomens from their hotter thoraces in flight with exchangers in their narrow, wasp-waist petioles (Heinrich 1996). Most of these exchangers can be bypassed by opening shunting vessels, so an animal can make an appendage dissipate heat instead when (usually during locomotion) needed.

All these countercurrent exchangers operate as steady-state devices. An unsteady version that briefly stores heat in both mammals and birds is known as well, again a scheme whose wide use was evident only after recognition of the first. Here the nasal passages of a North American desert rodent, the kangaroo rat, provided the initial case. Jackson and Schmidt-Nielsen (1964) showed that during exhalation heat moved from the air stream to the walls of the passages, so air left an animal not at body temperature but near ambient temperature, and in a dry atmosphere, slightly below ambient. Then, during inhalation, heat moved from passage walls to air, warming it and cooling the walls. In desert rodents the primary function of this heat exchanger appears to be water conservation, with over 50 percent of respiratory water loss (their principal mode of leakage) avoided by this condensation during exhalation and revaporization during inhalation. But they economize on heat as well, in amounts significant relative to overall metabolic rates—they recapture over 60 percent of the energy needed to heat and humidify inhaled air (Schmidt-Nielsen 1972). Camels conserve water in the same manner with their enormously surface-endowed nasal turbinates. But for them the concomitant thermal economizing may be detrimental rather than advantageous (Schmidt-Nielsen 1981).

Both children and adult humans exhale air close to core temperature. I wonder, though, about neonatal humans. My son, when about a week old (a smaller-than-average baby who is now a larger-than-average adult), seemed to be exhaling air that was quite a lot cooler than what came out of my own nose.

Buffering Fluctuations through Short-Term Storage

We may be less sensitive to the problems of temperature variation than of inopportune temperatures. Our large size buffers us from changes in ambient temperature and radiant regime. What constitutes a temporal extreme, though, depends on size. As large creatures we can ignore most events that last only seconds and need not take seriously most minute-scale phenomena. I can move a finger through a candle flame without discomfort, much less injury; and I once watched students on a Canadian campus going without coats from one building to an adjacent one at about −30°C. At the same time, few, even well bundled, waited in the open for buses.

So we encounter yet another problem of scale, that ever-lurking consideration in each of these chapters. Our perceptual world remains distant from that of a small marine snail caught on a large rock in summer sunlight at low tide or that of a thin sun-lit leaf on a tree when the normally ubiquitous air movements briefly abate. And, as Denny and Gaines (2000) remind us, the distribution of organisms more likely reflects local extremes, particularly temporal ones, than it does regional averages.

While for large animals minute-to-minute thermal fluctuations may not matter, variations on scales of hours clearly do. A particularly interesting case is that of a camel in a hot desert, faced with problems of both too much heat and too little water, something investigated by Schmidt-Nielsen et al. (1957) and later put into a general context (Schmidt-Nielsen 1964). Comparison of normal and shorn animals showed that fur reduces both heat gain and evaporative water loss. Beyond using fur, camels (dromedaries in North Africa) take peculiar advantage of the predictability of the main temporal fluctuation that affects them. When their access to water is limited, they permit their core temperatures to rise from about 34° to 40° during the day, secure in the knowledge that night will follow with cooler air and (usually) an open sky. They thereby reduce evaporative water loss (less sweating, mainly) almost threefold and halve overall water loss.

One might expect that only large creatures can play this particular game—a few large mammals, perhaps some of the more massive cacti (Nobel 2005 calculated time constants for the latter of several hours). Remarkably, at least one group of small desert succulent heats slowly enough to do so as well. These so-called stone plants (*Lithops* spp.) live largely buried in the soil of the Namib desert and the Karoo scrubland of South Africa. They protrude only about 2 millimeters above the surface but extend downward about 30 millimeters, as in plate 5.1. A translucent window on the top of each of the paired leaf-analogs admits light into the interior, with the photosynthetic tissue (the chlorenchyma) lining the bottom somewhat as our retinas line the inner rear surfaces of our eyeballs. Turner and Picker (1993) found that air temperatures ranged from 12° to 46° over the course of the day. So, very nearly, did plant surface, plant interior, and the surrounding soil a centimeter below the surface—all rose rapidly though the morning, peaked in the afternoon, and slowly dropped through the night. That's a lot more variation than experienced by a camel, but direct solar exposure without coupling to the surrounding soil would make matters worse—plants surrounded by Styrofoam insulation became considerably hotter than those in full contact with the soil.

So by combining its thermal mass with that of the surrounding soil, *Lithops* buffers its daily temperature changes and, critically, reduces its peak daytime temperatures. In addition, it takes advantage of the steep vertical thermal gradient in the soil. It couples, not to the hotter surface but to the cooler soil a short distance beneath, and then puts its most metabolically active tissues well down from that hot surface.

To emphasize the connection between the size of the system and the relevant temporal scale of fluctuations, we might return to broad leaves during periods of what we think of as still air. Convection, whose magnitude depends strongly on air speed, provides a major avenue of heat transfer. The speed of "still air" fluctuates rapidly and continuously, the result of passing turbulent structures and local convection. And with lots of surface and little volume, leaves are effectively small and have rapid thermal responses.

Some years ago I tried to get a sense of a leaf's thermal situation on a still, sunny summer afternoon by mounting a model leaf near the top of the forest canopy. The model, of cellulose acetate with black ink dots, had both the shape and thickness of a sun leaf of white oak (*Quercus alba*)—testing in the laboratory assured me that both its absorptivity and time constant came close to the values for real leaves. A tiny bead thermistor glued to its lower surface monitored mid-blade temperature, while a heated thermistor tracked adjacent air movement. Figure 5.2 shows a typical pair of tracks. Air temperature remained almost constant, and model temperature remained above it except when a cloud covered the sun. Most notably, the temperature of the leaf model was anything but constant; when the wind dropped, it rose with only a short lag. One rarely thinks of leaf temperature as such a wildly fluctuating variable; once alerted, one wonders about the biochemical implications of its rapid and continuous change.

Nobel (2005) calculated a time constant below 20 seconds for a broad leaf, quite consistent with the data from my model in figure 5.2. As one can see from the figure, so rapidly does air speed change that even a modest increase in time constant would yield significant thermal buffering. Thus, improved protection against temperature extremes would require vastly less mass than a camel or stone-plant-plus-soil. So increased thickness might well constitute a specific adaptation to assure lower peak temperatures during brief episodes of especially low wind—as opposed

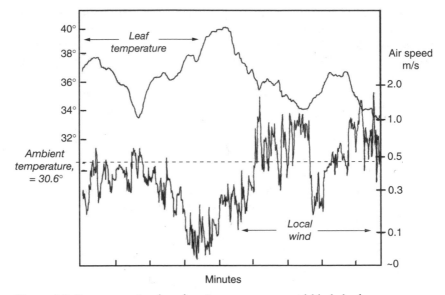

Figure 5.2. Representative data for air temperature, mid-blade leaf temperature, and wind speed for a model sun leaf of white oak, *Quercus alba*, on a typically windless summer afternoon in the Carolina piedmont. Note the nonlinear scale for wind speed.

to an incidental consequence of some other functional demand. Plants with small, thick leaves have long been termed "xerophytes" for their prevalence in dry habitats; the leaf structure is then "xeromorphic." Perhaps the plants might more appropriately be called "thermophytes," the lack of local water for evaporative cooling simply contributing to their underlying thermal challenge.

A functional explanation that focuses primarily on heat and only secondarily on water might explain the peculiar prevalence of plants with xeromorphic leaves in some well-watered places such as the swampy bogs of eastern North America. Traditional explanations invoke some kind of physiological dryness or deficiencies of nitrogen or calcium. But the results of a comparative morphological study by Philpott (1956) are consistent with a thermal rationale. She matched leaves of nineteen species from forest-surrounded bogs in Carolina (called "pocosins" in the region) with those of fourteen related plants from the Appalachian mountains directly inland. Whether looking at specific genera or at averages, the bog plants had smaller and thicker leaves. Small size would give better convective coupling to the surrounding air and therefore less deviation from ambient temperature; thicker leaves would heat more slowly during lulls. Thus the low wind and high humidity that makes these bogs notoriously unpleasant for people may just be major factors that challenge the local plants as well.

Some work by Kincaid (1976) provides more direct evidence that leaves may decrease size and increase thickness to lower peak temperatures through short-term heat storage. He collected holly (*Ilex*) leaves of a variety of species that experience different thermal extremes and exposed them to a wide variety of regimes in a very low speed wind tunnel in my laboratory. Among other manipulations, he subjected radiantly heated leaves to pulses of moving air, alternating 10 seconds of still air (< 0.01 meters per second) with 10 seconds of winds of 0.1 and 0.5 meters per second, conditions of light and air movement in a range they might encounter on a hot, windless day. Larger and thinner leaves heated significantly faster and further during lulls than did smaller and thicker ones. Indeed, the variation in behavior among the different species in the wind tunnel correlated satisfyingly with estimates of the importance of short-term heat storage from field data.

THE POSSIBILITY OF COUNTERCONVECTION

In examining how the physical world affects the adaptations and aspirations of organisms, I've been trying for general perspectives rather than conventional reviews. So I want to consider physical devices as yet unknown in living systems. After all, as an alternative to our normal search for functional explanations of specific features of organisms, one can look for organisms that use some hypothetical but plausible device. Per Scholander's recognition of the commonness, diversity, and general function of biological countercurrent exchangers—as well as much else he did—certainly shows the utility of the approach. In a sense, he played Hamlet to us Horatios; as Shakespeare put it, "There are more things in heaven and earth, Horatio, than are dreamt of in your philosophy."

Consider a scheme not yet known in a biological system. Countercurrent devices combine convective with conductive or diffusive transfer—fluid moves axially through pipes while heat or molecules conduct or diffuse laterally through the fluid and across the walls between the pipes. These two modes of moving heat or molecules can be combined in another way. This one could drive heat transfer below what we've been treating as a baseline, pure conduction, achieving, in effect, perfect insulation. Credit for asking about its possible roles should go to an engineer, the late Lloyd Trefethen. But when he described the scheme and asked me whether it had biological examples, I could offer no specific instance. Perhaps some reviewer or reader will recognize a case of what has been called "counterconvection." It operates in the following way. (While we'll focus on heat transfer, bear in mind that diffusive material transfer and a concentration gradient could replace heat conduction and a temperature gradient.)

Imagine a porous, conductive barrier, one through which both fluid and heat can pass, between two compartments that differ in temperature, as in figure 5.3. Heat conducts from warmer to cooler side. That conduction, though, is exactly offset by fluid forced through the barrier. In that way, all the heat that would otherwise be conducted down the thermal gradient gets transferred to fluid flowing up that gradient. And fluid flowing up the thermal gradient, now preheated, no longer cools the warmer compartment as it enters. So heat moves down a thermal gradient while fluid moves down a pressure gradient, balancing conduction in one direction by convection in the other. Balance will be achieved when

$$\frac{k}{x} = v\rho C_p, \qquad (5.2)$$

where k is thermal conductivity, x is the thickness of the barrier, v is flow speed, ρ is fluid density, and C_p is heat capacity (or specific heat at constant pressure).

The principal difficulty, to provoke proper skepticism at the start, is that the mechanism does not (at least as I see it) lend itself to operation as a closed cycle. Fluid will accumulate in one compartment, so draining it in any ordinary way will most likely offset anything gained. Actively pumping fluid will leave the system still worse. This suggests looking at systems in which fluid ordinarily enters or leaves and can be secondarily pressed into counterconvective service or else on systems that operate only part time, perhaps during periods of particular environmental stress.

Does the possibility pass quantitative muster? Consider two cases in which hypothetical organisms find themselves in a dangerously hot situation:

1. A spherical animal with 1 square meter of outer surface (one 0.56 meters in diameter) and an insulating layer of fat 0.01 meters thick is exposed to an outside temperature 10° above body temperature. High humidity or a liquid external medium prevents evaporative heat transfer. If fat's conductivity is 0.21 waters per meter-degree, Fourier's law for conduction predicts heat entry at 210 watts. Expelling it in the form of water, with a heat capacity of 4.2 kilojoules per kilogram-degree, would take only 5 milliliters per second. Still, that amounts to 18 liters per hour, which would use up the entire volume of the animal in just a little over 5 hours, making the scheme an unattractive long-term fix. That 210-watt heat entry would normally cause the animal to heat (initially at least) at about 1.9 degrees per hour, which ought to be tolerable for short periods.

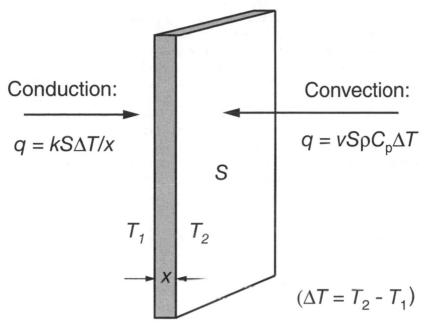

Figure 5.3. Heat conduction, left to right, and convection, right to left, in a counterconvective arrangement. S and x are slab area and thickness respectively, k is thermal conductivity, v is cross-slab flow speed, ρ is fluid density, and C_p is the heat capacity of the fluid.

So counterconvection would not work well for long periods and would be unnecessary for short periods. Still, the scheme isn't so implausible that it can be dismissed for all scales of size, time, and temperature.

2. Another spherical animal of the same size and faced with the same temperature difference has no insulating fat; instead it has a fur coat of the same (0.01 meter) thickness. Heat conducts inward through the fur while perspired liquid water moves outward and then obligingly disappears with no additional thermal consequences. Conductivity is now 0.025 watts per meter-degree, and heat will enter at 25 watts. That requires an outward flow of water of only 0.6 milliliters per second or 2.15 liters per hour. The animal thus contains about a 44-hour supply, enough, one might guess, to deal with a hot afternoon in the secure knowledge that night will follow in a few hours. But one further calculation puts this sanguine scenario as well in a less favorable light. With water's heat of vaporization of 2.44 megajoules per kilogram, dealing with an input of 25 watts by evaporative cooling would take only 0.037 liters per hour,

about sixty times less. Thus the scheme, while possible, makes sense only where evaporative cooling cannot be relied upon.

What should we conclude? While we should not dismiss the possibility of counterconvection, the requirements for it to be worthwhile turn out to be daunting. Still, conduction through a material of low conductivity and flow through a porous barrier, the requirements for it to happen, are biologically ordinary. One can produce enough bulk flow through such a barrier with only a modest pressure gradient, and organisms often either absorb or excrete liquid water for other purposes at appropriate rates. Someone with greater insight, creativity, or luck might just find a case—once alerted to the possibility.

CHAPTER 6

Gravity and Life in the Air

MAKING LIGHT OF WEIGHT

In our perceptual world, no physical agency imposes itself with greater immediacy than does gravity. We depend on it to walk or run; it injures us if we trip. It makes each of us about half a centimeter shorter at the end of each day than when we first arise. Our flesh sags as we age; more slowly, glaciers move downhill and rivers form deltas. We dream of escaping its constant crush, although our recent experiences in orbiting spacecraft reveal the difficulty of opting out of its perpetual pervasiveness. Physicists may regard the gravitational attraction between two objects as the universe's definitional weak force, but to us large, terrestrial creatures it feels anything but weak.

Since the consequences of gravity depend on one's size, scaling will loom at least as large in this and then the next as in any of the preceding chapters. Even more important than the ways gravity's effects scale will be another message—the surprisingly wide range of biological situations in which it plays some role. We know that no massless world exists; in fact a weightless world is almost as hard to imagine.

That contrast, mass versus weight, needs a few words. Newtonian mechanics lumped two distinct kinds of mass, inertial mass and gravitational mass. Only early in the twentieth century was the physical basis of their apparent equality finally revealed—for particulars, one should consult a physicist; I'll worry only about the difference. An inertial mass resists acceleration, as expressed in Newton's first and second laws; quantitatively, mass equals force divided by acceleration—the familiar $F=ma$. A gravitational mass attracts any other mass. It exerts a force equal to the product of the two masses, divided by the square of the distance between them, times a universal gravitational constant, in proper SI units, 6.67×10^{-11} newton-square meters per square kilogram. (One should avoid using "gravitational constant" for the acceleration of gravity at the surface of the earth, commonly designated g.) In our world that other mass is that of the earth itself, 5.976×10^{24} kilograms, and the basic distance is that from the earth's surface to its center of mass, 6370 kilometers. These data give $g=9.8$ meters per second-squared. Weight follows from this second manifestation of mass; with $F=mg$ we can then convert mass (kilograms) to terrestrial weight (newtons).

Our lack of intuitive feeling for the difference between mass and weight just reflects inexperience with situations in which the constant of proportionality differs from the terrestrial 9.8 meters per second-squared. Think back to video images of astronauts ambulating on the surface of the moon. Unsurprisingly, they adopted a rather bouncy gait when going straight ahead; one could easily fail to notice the greater forward tilt needed to get going. Turning, though, looked distinctly strange, in particular the way it took a greater lean to change direction. These greater tilts for starting, stopping, and turning provided sufficient force so weight-deprived individuals could accelerate their unchanged masses, so they could apply what weight they retained with sufficient effect. As it happens, the human neuromuscular machine copes remarkably well with this completely novel sixfold reduction in weight.

THE OTHER FORCES THAT MATTER

Life contends with lots of forces besides that of gravity. In most situations, though, we only need worry about one or two forces. The first item in an analysis commonly consists of identifying the forces at work and their relative importance. An extreme example from my own experience should emphasize the point.

Some time ago, I was asked to evaluate a claim that the asymmetries of the mammalian body, in particular of our own, could be traced to effects of the Coriolis force from the earth's rotation on the evolution of terrestrial animals. That force (really a pseudoforce, like the so-called centrifugal force) results from the rotating earth's spherical rather than cylindrical shape. Thus (as in figure 6.1) an object in the northern hemisphere moving north must move inward toward the earth's axis of rotation as well. Angular momentum being conserved, it should rotate faster. The effect will be felt as an eastward force, a force to the right of its path, since the earth rotates from west to east. When moving south, the object will move outward and as a result try to rotate more slowly—an effect felt as a westward force, but still to the right of its path. Clearly a slightly sturdier right leg ought to confer an advantage, putting an animal, one might say, a leg up. The same argument was applied most ingeniously to our many other anatomical asymmetries.

But I found against the plaintiff, so to speak, making my case by comparing the magnitude of the Coriolis force with that of the gravitational force. The former is twice the product of the object's mass (m), the speed at which the object moves north or south (v), the earth's angular velocity

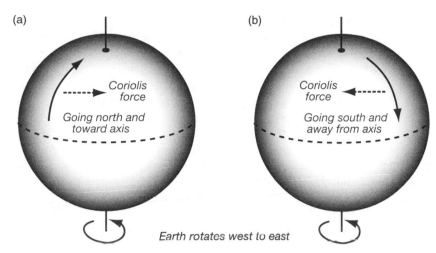

Figure 6.1. The origin of the Coriolis force for something moving (a) northward and (b) southward, in the northern hemisphere. While the force may be eastward and then westward relative to the earth, it remains to the right with respect to the mover.

(Ω), and the sine of the latitude (ϑ). Mass times gravitational acceleration (mg) gives the gravitational force. In their ratio, mass cancels, and we get

$$\frac{F_c}{F_g} = \frac{2v \cdot \Omega \cdot \sin \vartheta}{g}.$$

(6.1)

A most favorable possible scenario might consider an animal living at 45° latitude and spending its life going north or south at 1 meter per second. Under these conditions the ratio will be 1:100,000. To me that seemed to offer evolution precious little advantage with which to work; for more evolutionarily reasonable lower speeds and latitudes, the ratio would be still less auspicious. In short, little about our persons can be attributed to the Coriolis force—however dramatic its effects on, for instance, weather patterns.

[As a reader of the manuscript pointed out, the equation should not be applied in this simple form to the large bodies of air responsible for our weather. It ignores buoyancy, tacitly assuming that the density of the mass at issue far exceeds that of the atmosphere. Incidentally, Persson (1998) provides an engaging introduction to Gaspard Gustave de Coriolis (1792–1843) and his force.]

While the terrestrial Coriolis force may have no direct relevance to organisms, many other forces must matter in addition to that of gravity. Hydrostatic and aerostatic forces squeeze or expand organisms. Tensile, compressive, and shearing forces variously distort their shapes. The viscosity and dynamic pressures of flows impose both drag and lift. The inertia of fluids can exert major transient forces, as when the surface of a body of water is slapped—by a hand or, more significantly, when a basilisk lizard runs across a stream (Glasheen and McMahon 1996). Transpiring trees as well as water striders depend on surface tension. And so on. What most often determines the practical consequences of gravity is its magnitude relative to the other forces at work.

The engineering community, the fluid mechanists in particular, have long used a variety of dimensionless ratios of one force to another, analogous to that of equation (6.1), to evaluate their relative importance. Gravitational force contributes to many of them, either as numerator or denominator depending on the prejudice of the particular field in which the ratio originated—which force carried the load and which constituted a nuisance. For instance, the Bond number, below, has mainly been used for gravity-driven flows with interfaces in porous media, so gravity makes the system go, while surface tension acts as a brake. With gravitational force as numerator and surface tension as denominator, high rather than low values are Good Things. If, instead, the ratio had been contrived by a biologist concerned with animals supported atop a pond by surface tension, gravitational force would have dropped to the denominator.

Among the dimensionless ratios that include gravitational force (from Weast et al. 1987; new editions of the *Handbook of Chemistry and Physics* no longer give dimensionless ratios):

(i) Bond number—as mentioned, gravitational force to surface tension force:

$$Bo = \frac{(\rho_o - \rho_m)l^2 g}{\gamma}. \qquad (6.2)$$

ρ_o and ρ_m are the densities of object and liquid medium respectively, l is a characteristic length of the system, the choice depending on the particular phenomenon at hand, and γ is surface tension. (An analogous expression goes as the Baudoin number, Ba, cited in a recent and biologically relevant paper by Hu et al. 2007.)

(ii) Froude number—inertial force to gravitational force:

$$Fr = \frac{v^2}{gl}. \qquad (6.3)$$

Choice of l, again, depends on the system.

(iii) Bagnold number—drag to gravitational force:

$$Ba = \frac{3C_d \rho_m v^2}{4d\rho_o g}. \tag{6.4}$$

C_d is the object's drag coefficient and d its diameter. It resembles the Froude number because the underlying formula for drag [1/2 $C_d \rho_m S v^2$, with S for projecting area normal to flow, as in equation (6.12), below] tacitly presumes that drag is an inertial force and so ignores viscous, buoyant, and other sometimes contributory forces.

(iv) Grashof number—buoyant force to viscous force:

$$Gr = \frac{\rho^2 g \beta (\Delta T) l^3}{\mu^2}. \tag{6.5}$$

β is the coefficient of thermal expansion of the fluid, ΔT the temperature difference, and μ the viscosity. Grashof number appeared previously in chapter 4.

(v) Galileo number—gravitational force to viscous force:

$$Ga = \frac{l^3 g \rho^2}{\mu^2}. \tag{6.6}$$

Since buoyancy amounts to a manifestation of gravity, similarity between this one and its predecessor should be no surprise. Both the Grashof and Galileo numbers, as well as a few others, include as a factor the Reynolds number,

$$Re = \frac{\rho l v}{\mu}, \tag{6.7}$$

the ratio of inertial to viscous force—thus density and viscosity appear as second powers in both.

In all these dimensionless numbers (as well as others), the larger the system, the more important gravity becomes relative to other forces. Whether g appears as numerator or denominator, some size factor appears with it. I can think of no exception to that rule, although I hesitate to assert its universality. Nothing has broader relevance for scaling.

GOING UP AND DOWN

Besides keeping our atmosphere from drifting away, gravity makes its outer portions squeeze down its inner portions, so a pressure increase accompanies an approach to the earth's surface. (Only a tiny part of that increase comes from the increase in gravitational force as the earth's center is approached.) As in any ordinary gas mixture, atmospheric density follows pressure—some consequences of altitude change result from density change, others from pressure change. In particular, pressure affects the solubility of gases in liquids. A carrier of respiratory gas such as hemoglobin, suitable for reversible binding with oxygen at one altitude, will not work as well at a very different one. Mammals adapted by ancestry (as opposed to individual experience) to high altitudes have hemoglobin variants with greater affinities for oxygen (Hall et al. 1936).

The volume of a helium- or hydrogen-filled balloon increases as it rises; if its buoyancy (upward force) varies with volume and its drag (downward force) with surface area, its ascent speed will gradually increase. Organisms, though, do not use buoyant bags to ascend in air. But the volume increase still matters, requiring internal air containers either to be surrounded by stretchy walls or to be vented to the outside. We vent our middle ears into our respiratory passages through a pair of Eustachian tubes, and ascents and descents in aircraft or elevators with plugged tubes cause pain and temporary auditory impairment. Birds, facing the problem in more severe form, vent all their air-filled bones.

The external effects of that volume increase with altitude (or with anything else that lowers pressure) may be more important. If a patch of ground heats more than the surrounding area, the locally warmer air above it may rise. It initially forms a column, then a round bubble, and finally a torus, a tubular vortex with its ends connected into a doughnut. That rising torus, typically over a highway or plowed field, can provide an elevator for pollen, spores, seeds, and small organisms.

One might expect that any ascent will be brief, since such relatively dense objects will descend relative to the local air and soon fall out of the ascending torus. That need not be the case—the torus forms because air at the periphery of the bubble is slowed by the surrounding air. As a result, air near the periphery descends relative to the overall structure, and air near the inner portion of the ring must rise. So something near the inside margin of the toroidal ring (next to the doughnut's hole) can fall steadily relative to the surrounding air and avoid falling out. Many birds appear to do just that, soaring in circles whose radii are smaller than the radii of the tori. That need to glide in fairly tight circles has been invoked to explain the typically shorter and broader wings of terrestrial—by contrast with marine—soarers. Other creatures such as tiny spiders and insects may exploit the same opportunity by paying out long threads that partly enwrap these vortices, although we lack specific documentation.

Locally warmed air rises; locally cooled air should fall. Such cold, downslope currents, mainly at night, often occur in mountainous terrain. Extreme versions go by names such as "air avalanches," "mistrals," and "williwaws" and may reach 40 meters per second. Geiger (1965) describes ones of surprisingly regular short-term periodicity.

Large-scale mixing may not always suffice to keep atmospheric composition uniform, and the resulting local changes in atmospheric density can have serious physical and biological consequences. Local enrichment with a light gas such as methane will produce upward bubbles, columns, and toroids just as does local heating, if on a smaller scale and less portentously. Local enrichment with a heavy gas will, in the absence of significant wind, lead to a stable, enriched layer at ground level. A ground-level layer of carbon dioxide, not normally regarded as a serious toxin, led to the Lake Nyos disaster of 1986 in West Africa (Kling et al. 1987). After the outgassing that began at some supersaturated spot deep within the lake, CO_2 reached sufficient concentration to cause the immediate death of the 1,700 or so people of the surrounding villages.

FALLING IN AIR

When acting directly on the mass of an organism, gravity has consequences both more common and serious than anything resulting from changes in atmospheric pressure and density. Unrestrained, a body accelerates downward at 9.8 meters per second-squared, so

$$v = \sqrt{19.6d} \qquad (6.8)$$

The impact speed after a fall of a meter will be 4.4 meters per second, tolerable for most organisms under most circumstances. A 10-meter fall should give an impact speed of 14 meters per second and a 100-meter fall of 44 meters per second, both ordinarily hazardous.

In reality a body falling in air accelerates ever more gently, asymptotically approaching a state where the downward gravitational force equals its upward drag, where the Bagnold number (above) approaches 1.0. The apparent calculational simplicity, though, turns out to be deceptive. As noted when considering trajectories in chapter 2, C_d, the drag coefficient, varies peculiarly. For very small things falling in air—fog droplets, pollen grains, and so forth—it varies inversely with speed (more specifically, with the Reynolds number and as described by Stokes' law), while for large, fast, dense things it remains very nearly constant. In between, from ejected spore clusters, falling seeds, and small, flying insects to medium-sized flying birds—no biologically trivial realm—we run into several shape and Reynolds-number-dependent transitions, some abrupt and some gradual (Vogel 1994b).

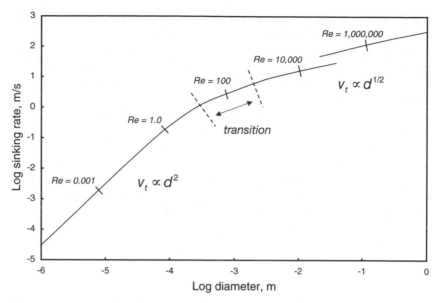

Figure 6.2. Terminal sinking rates of spheres without induced internal motion and of the density of water as a discontinuous function of Reynolds number (eq. 6.7, using the density and viscosity of air).

Figure 6.2 gives terminal speeds for spheres of the density of water falling in air of sea-level density, calculated using the formulas of chapter 2 and an interactive program. (Streamlined bodies, at least ones of large or moderate size, reach higher terminal speeds, while irregularly shaped or tumbling bodies descend more slowly.) Bigger inevitably means faster, but for small spheres terminal speed is especially size-dependent, increasing with the square of diameter according to Stokes' law, while for large ones terminal speed increases with only the square root of diameter.

Nonetheless one should not assume that drag need always be taken into account. In figure 6.3 the same equations have been used to view the approach to terminal speed for spheres of a range of sizes (again at one atmosphere and of water's density). Large bodies need long drops to approach terminal speeds, while small ones get there so shortly after release that one can assume instantaneous acceleration. For instance, to get within 5 percent of terminal speed, a 100-millimeter sphere needs about 100,000 millimeters of fall; a 10-millimeter sphere needs about 2000 millimeters of fall; a 1-millimeter sphere needs about 100 millimeters of fall; a 0.1-millimeter sphere needs about 0.1 millimeter of fall; a 10-micrometer sphere needs a mere 0.01-micrometers of fall. A 10^4-fold decrease in diam-

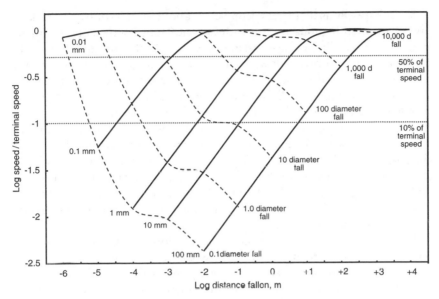

Figure 6.3. How the approach to terminal velocity varies with the size of spheres of density 1000 kilograms per cubic meter falling in air.

eter yields a 10^{10}-fold decrease in the dropping distance to get to 95 percent of terminal speed.

Alternatively, one can view (as in figure 6.3, again) the scaling in terms of the diameters of the spheres, especially handy since falling distance over diameter is a nicely dimensionless "length." To get to our benchmark 95 percent of terminal speed, the 100-millimeter sphere takes 1,000 diameters, the 10-millimeter sphere takes 200 diameters, the 1-millimeter sphere takes 100 diameters, the 0.1-millimeter sphere takes 1 diameter, while the 10-micrometer sphere takes a mere 0.001 diameter. (That peculiarly irregular variation of drag with speed for falling spheres and most other ordinary shapes makes the sequence somewhat erratic.)

When falling in air, drag and terminal speed carry biological significance mainly for small objects—unless one considers selectively questionable behavior such as sky-diving by humans. Still, even small ones may run out of range of Stokes' law, reliable only for Reynolds numbers decently below 1.0. So one may need to do iterative calculations such as those that generated figures 6.2 and 6.3. A sphere 1 millimeter in diameter will exceed $Re = 1.0$ after a fall of less than a millimeter, long before reaching terminal speed (at which $Re = 230$), indeed, well before drag has begun to alter its motion much at all. Even a 0.1-millimeter sphere, about at our visual threshold, will exceed $Re = 1.0$ at terminal speed.

What happens as diameters drop below about 10 micrometers? Stokes' law gives unambiguous results with no uncertain coefficients. Unfortunately an exceedingly basic assumption, one only infrequently made explicit, begins to break down and as a result to restrict its reliability. For the most part the field of fluid mechanics treats fluids as continua—nonparticulate, infinitely divisible without loss of character. While 10 micrometers remains far above molecular dimensions, molecular phenomena do start to intrude. What's happening is that, as their diameters approach the mean free paths of the molecules of the surrounding fluid, the terminal speeds of particles begin to disobey the law. For atmospheric air, mean free paths are of the order of 0.1 micrometers. That's only two orders of magnitude smaller than the 10-micrometer spheres considered here and closer still to, for instance, a 4-micrometer spore of the fungus *Lycoperdon* (Ingold 1971).

In effect, the Brownian motion due to random collisions with moving gas molecules rises to the same scale as that caused by gravity, so motions become irregular, eventually having only a statistically downward bias. The relative magnitude of the effect increases rapidly with decreasing particle size both because gravitationally driven descent speed decreases and because the effective Brownian displacement speed increases. (The latter, as noted in chapter 1, is a peculiarly duration-dependent speed, here the square root of the quotient of twice the diffusion coefficient for a particle of a particular size divided by a reference time.) For a 10-micrometer particle (still of water's density sinking through air) and a reference time of 1 second, Brownian displacement speed is less than a thousandth of gravitational speed; for a 1-micrometer particle (perhaps an airborne bacterium), Brownian displacement speed rises to a fifth of gravitational speed; for a 0.1-micrometer particle, Brownian speed approaches a hundred times gravitational speed (Monteith and Unsworth 1990; Denny 1993).

Further clouds on the horizon need more mention than they usually get, suggesting caution in adopting textbook equations. The equations that generated figure 6.3 assume quasi-steady motion in an unbounded fluid; that is, they take no account of any special phenomena associated with acceleration. At least two unsteady phenomena can take on importance, one largely independent of the fluid's viscosity, the other its direct consequence.

First, when a body accelerates in one direction, fluid must accelerate in the other. The latter requires force no less than the former; it goes as the "acceleration reaction force" or simply the "acceleration reaction." For a sphere, one calculates the extra force by presuming that the body has an additional mass equal to half that of the volume of fluid it displaces—

that 0.5 is the "added mass coefficient" of a sphere (Daniel 1984; Denny 1988). So real accelerations are less than what one calculates for a quasi-steady case—as if drag were increased, but with the effect scaling with volume rather than diameter or cross-section. Decelerations are also re-duced (values of deceleration, the opposite of distances needed!), with the acceleration reaction now opposing drag. For a sphere of biologically relevant density in air, the acceleration reaction will usually be negligible next to drag. It should matter, though, for a buoyant balloon just after release. In water, the acceleration reaction can be a major factor. In at least one circumstance it dominates—for the initial ascent of a bubble of gas in a liquid. Here the mass, even half the mass, of the displaced fluid far exceeds that mass of the accelerating body, so neglecting the accelera-tion reaction overestimates acceleration by several orders of magnitude (Birkhoff 1960).

Second, setting up a steady-state flow pattern around a body takes time, so a history term may be significant during acceleration. Again, accelera-tions are reduced, here because velocity gradients and thus viscosity-generated shear forces are more severe than otherwise expected. Again, the effect, often called the "Bassett term" (Michaelides 1997; Koehl et al. 2003), will only rarely be important for ordinary bodies accelerating in air.

[The Bassett term is analogous to the long-known Wagner effect (Wag-ner 1925; Dudley 2000), a delay in the development of aerodynamic lift as an airfoil begins to move. Moving those initial seven or eight wing widths cannot greatly tax the taxiing to take-off of an airplane; but it de-mands special devices for animals that, lacking rotational propellers, must flap their wings and thus start each wing twice for each stroke.]

Nor do the usual equations worry about wall effects or interparticle interactions, which occur whether a body is accelerating or moving steadily. Like the Bassett term, they result mainly from viscosity and the resulting velocity gradients. A body falling near a wall falls more slowly, some-times much more slowly; the lower the Reynolds number, the more se-vere the effect and the more distant a confounding wall can be. And one starting from a surface will have a lower initial acceleration. Conversely, a body falling in the wake of another will experience lower drag and tend to catch up; a cloud of tiny bodies can thus coalesce as the bodies fall. These effects are more likely to be of importance in air than are the ac-celeration reaction and the Bassett term; at the same time they are easier to identify and avoid. Not that biologists do so consistently. A substan-tial literature for sinking rates in nature reports measurements of the de-scent speeds of clouds of individuals (and so too fast) in worrisomely narrow tubes (and so too slow).

ANOTHER WAY TO DESCEND SLOWLY

The higher a body's drag, the slower gravity will make it descend, at least after drag has acted for a while. Alternatively, descents can be slowed with some kind of lift-generating airfoil—a device that produces a force component at right angles to the oncoming airflow in addition to the inescapable drag, acting parallel to flow. Not only does this mode of descent-slowing have quite a different aerodynamic basis, but it imposes an antithetical requirement. The effectiveness of a lift-producing airfoil depends on its lift-to-drag ratio. That implies minimization rather than maximization of drag.

At higher Reynolds numbers airfoils can achieve greater lift-to-drag ratios, so slowing descents with lift becomes increasingly attractive as systems enlarge. Thus airborne seeds (or fruits or seed-leaves—more generally, "diaspores" or "propagules") that slow descents by increasing drag mostly have masses below 50 to 100 milligrams; very few of what Augspurger (1986) terms "floaters" exceed that benchmark. Most heavier ones, such as the samaras of maples, ashes, and tulip poplars, employ lift-producing airfoils. Conversely, samaras come no lighter than about 10 milligrams (Azuma 1992). By contrast, animals such as flying squirrels that can control their aerial postures blur the boundary, with no hard and fast distinction between drag-based parachuting and lift-based gliding—more about these shortly.

What the lift-to-drag ratio sets is the angle with which a gliding airfoil descends, whether it's a glider moving in one direction or a samara autogyrating downward along a helical path. Specifically, the lift-to-drag ratio, L/D (for the entire craft if made of more than a single airfoil) equals the cotangent of the angle relative to the horizontal, ε, of a steady-speed descent:

$$\frac{L}{D} = \cot\varepsilon \quad \text{and} \quad \frac{D}{L} = \tan\varepsilon. \tag{6.9}$$

But although it sets the path, the ratio doesn't fully determine the speed at which the craft approaches the earth. For a steady glide, descent speed depends as much on the amount of upward force needed, which must equal the weight of the craft. Since lift (like drag) is proportional (putting aside some secondary matters) to the square of speed, that square of speed has to vary directly with weight. In other words, doubling the weight of a glider increases its steady-state speed (both overall and descent speeds) by 1.414. Thus in still air (and assuming a unidirectional glide, not an autogyrating vertical descent), the heavier glider will go about as far when released from a given height, but it will get there faster.

That independence of glide angle and weight may underlie the large size of some fossil fliers, whether insects (Paleodictyoptera), reptiles (pterosaurs), or high extinct birds such as *Argentavis* (Chatterjee et al. 2007). But an up-front caution is needed. One might imagine that, assuming biologically ordinary tissue densities, an increase in the weight of a glider will be offset by the increase in wing area and thus lift, which varies directly with it. But the scaling of wing area and body mass, about which more in the next section, undercuts that offset. For isometric craft of constant density, weight, W, will vary with length cubed while lift will vary with wing area, S, and thus body area and length squared:

$$W/S \propto l. \tag{6.10}$$

That variable, W/S, goes by the name "wing loading." As a consequence of that length-proportionality, if size increases isometrically, bigger must mean faster. That demands some combination of more wind-dependent takeoffs and harder landings. Perhaps those large fossils tell us that back when few fast terrestrial predators lurked, isometry could be put aside—the increased fragility of light construction and disproportionately large wings may have been less disadvantageous. Or that head-wind-dependent takeoffs were more often tolerable.

(Wing loading may enjoy a weight of tradition, but it ignores at least one potentially confounding factor. Long, narrow wings do better than short, wide ones, something now well understood, but evident only empirically in the early days of flight. To avoid giving equal weight to length and width in wing area, an alternative variable, "span loading," the ratio of weight to the square of the wingspan, sometimes finds use. Choice of variables matters little for wings of ordinary proportions or for comparisons among wings of similar shapes.)

So both glide angle and wing loading (the relevant form of weight) enter the picture, and wing loading is strongly inimical to large, living gliders. By contrast, glide angle gets better (lower, as conventionally defined) with increasing size, so it's inimical to small craft. But it does so less sharply. For a high-quality airfoil, lift, or properly the lift coefficient (C_l), depends only slightly on Reynolds number, at least for *Re*'s above those of fruit flies, around 100; it usually has a value of about 1.0 or a bit less at a maximum lift-to-drag ratio. On the other hand, drag, expressed as the analogous drag coefficient (C_d), drops with increases in the Reynolds number. Formally defining those coefficients, we have

$$L = C_l \rho_m v^2 S/2 \quad \text{and} \tag{6.11}$$

$$D = C_d \rho_m v^2 S/2, \tag{6.12}$$

where ρ_m is the density of the medium and S is the projecting area, normal to flow, of the airfoil. In effect, the best achievable lift-to-drag ratio will increase (if complexly) with size, putting small gliders at a disadvantage when it comes to glide angle.

How much so? Since effective airfoil design itself varies with Re, we may be misled by looking at how a specific airfoil's or aircraft's performance varies with Reynolds number. So we're better off comparing a heterogeneous collection, in effect selecting ones that have proven effective under their individual operating conditions. Figure 6.4 gathers data for lift-to-drag ratios for airfoils as a function of Reynolds number. Bear in mind that these data come from diverse sources, with embedded estimates to render them commensurate. A competitive sailplane may have a ratio of 50; a top-flight bird, an albatross, of 20; the hindwing of a desert locust about 8; flies and bees around 2; test airfoils at $Re = 10$ and $Re = 1$ of 0.43 and 0.18 respectively. Clearly, small fliers cannot achieve glide angles as low as those of large fliers, and the high glide angles at Reynolds numbers below about 500 make gliding itself impractical.

In one sense, though, the inferior glide angles of insects (and, although less extreme, of birds) may mislead us. In that earlier assertion that as wing loading went up with body size so must flying speed lies a compensating advantage of small, if not very small, size. In the real world, gliding for greatest distance in a temporally and spatially uniform atmosphere represents both a worst and an uncommon case. We know quite a few ways gliders can take advantage of atmospheric structure, what we call "soaring" as opposed to simple gliding. Most schemes for soaring depend as much on time aloft as on the horizontality of simple gliding. Time aloft, of course, varies inversely with sinking speed, so time aloft is no worse for a flier that descends twice as steeply if it flies half as fast. With still-air time aloft as the criterion, gliding/soaring retains utility down into the large insect range—it may even improve. A limit line drawn through the upper left set of points in figure 6.4 has a slope of 0.217. Converting from lift-to-drag ratio to sinking speed tells us the latter will vary with $Re^{0.116}$ for isometric gliders of equal lift coefficients (eq. 6.11). So the smaller (lower Re) glider will approach the earth somewhat less rapidly in still air or, of more relevance, slower ascending air will suffice to keep it aloft. Tucker and Parrott (1970) make just this point, noting that soaring birds such as some vultures can achieve lower minimum sinking speeds than high-performance sailplanes.

Why don't tiny, even microscopic, gliders fill the skies? Flapping fliers blur the issue, with what appears to be a gradual diminution with decreasing size of the extent to which they employ intermittent gliding; locusts, butterflies, and dragonflies glide at least a little; bees and flies do not. Purely passive gliders provide a clearer dichotomy. As noted earlier, among

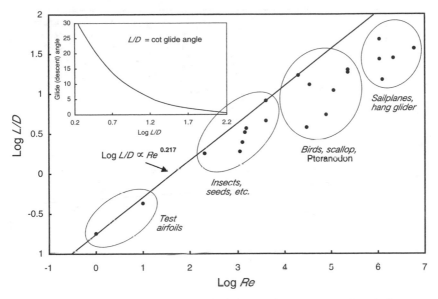

Figure 6.4. The scaling of the lift-to-drag ratio with Reynolds number, with an empirical limit line. The inset provides a conversion of the ratio to the steady-state glide angle with respect to the horizontal. Data for human-carrying craft from commercial websites; *Pteranodon* from Bramwell (1971); scallop from Hayami (1991); birds from Withers (1981); insects from sources in Vogel (1994b), p. 249; seeds from Azuma (1992); test airfoils from Thom and Swart (1940).

plants—those wind-dispersed, autogyrating samaras, mainly—gliders get no smaller than about 10 milligrams and fly at Reynolds numbers no lower than about 500. Among animals, purely passive gliders become uncommon below about 1 gram (McGuire and Dudley 2005). That initially puzzling two-order-of-magnitude difference may just be a matter of adaptational opportunities or lack of alternatives such as active flight.

Still, the absence of much of a fauna analogous to the samaras seems odd. Some tiny arboreal animals do glide—ants recently have been shown to do so (Yanoviak et al. 2005). One suspects that spiders, looking for lateral attachment points for webs within windless forests, might do likewise. Now that we know of gliding ants, perhaps someone will seek them out. But I doubt very much that the arthropods will provide examples of tiny gliders of the commonness of samaras. Part of the difference may turn on the respective roles of glide distance in linear versus autogyrating gliders. The samara cares only about time aloft, to which glide distance contributes only indirectly. But the ant needs to get from one tree-trunk to another, and a low *L/D* imposes a major altitudinal loss in the process.

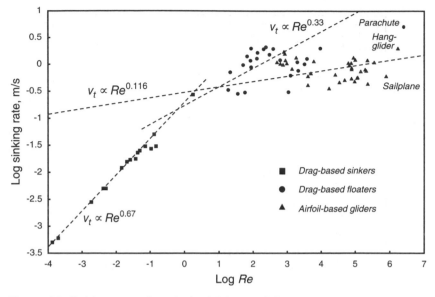

Figure 6.5. Sinking rates of passively sinking or gliding systems spanning an especially wide size range and anything but isometric. Note that while the exponents of the scaling lines can be justified, their positions are arbitrary. The exponent of 0.116 comes from the empirical limit line of figure 6.4; the exponents of 0.33 and 0.67 are theoretical and come from figure 6.2. Sources: Azuma (1992), Bramwell (1971), Gibo and Pallett (1979), Ingold (1971), Jensen (1956), McGahan (1973), Niklas (1984), Okubo and Levins (1989), Parrott (1970), Pennycuick (1960, 1971, 1982), Rabinowitz and Rapp (1981), Tennekes (1996), Trail et al. (2005), Tucker and Heine (1990), Tucker and Parrott (1970), Verkaar et al. (1983), Ward-Smith (1984), Werner and Platt (1976), and Yarwood and Hazen (1942).

The data collected in figure 6.5 may provide a bit of insight for the overall decrease in the use of gliding in smaller forms. Perhaps gliding's slight improvement in time aloft with decreasing size eventually competes with an alternative that offers still better scaling. The left, linear portion of figure 6.2 implies that sinking speed for objects retarded by drag will vary with $Re^{0.67}$, more drastic scaling than the $Re^{0.116}$ for lift-based retardation, an implication well confirmed by the real-world data of figure 6.5. Smaller becomes not a bit but a lot better than for craft that slow their descents by maximizing drag.

Drag-based floaters blur the contrast just a little. While generally lighter than the lifters, they operate in a realm with an intermediate scaling exponent—with sinking speed proportional to $Re^{0.33}$ as on the right side of figure 6.2. And without obvious exception, they keep sinking rates reasonable through drastic surface proliferation, equipping themselves with all manner of hairs, fluff, and appendages.

Still, too quick a look at such data for both glide angle and sinking speed may mislead us in another way. It accords poorly with the repeated evolution of remarkably bad gliders in several groups of terrestrial vertebrates—frogs that glide using oversize webbed feet, lizards that glide with lateral trunk extensions, squirrels and phalangers that glide with thin skin that stretches from fore to hind legs, even snakes that glide with a bit of body flattening and a cross-flow body orientation (see, for instance, Dudley and DeVries 1990; Norberg 1990). None of these achieves an especially high lift-to-drag ratio for its size; values run from a little over two to a little under five (Socha 2002). The apparent paradox may stem from the way both glide angle and descent speed tacitly refer to steady-state activities. These latter may be the exception rather than the norm in such animals—which is why I omitted them from figure 6.5. A typical trajectory consists of an initial outward and then downward leap, with only a minor aerodynamic component; as increasing speed generates significant lift the path becomes ever less vertical. Major and deliberate drag increases precede landings, in the process raising lift at the expense of speed and the lift-to-drag ratio—reinvesting the kinetic energy of the initial drop when airspeed is no longer an asset.

Recent work on flying snakes, genus *Chrysopelea* (Socha et al. 2005; see also www.flyingsnake.org), and on lizards, genus *Draco* (McGuire and Dudley 2005), provide graphic examples. Clearly the old and often quoted distinction between parachuting and gliding, whether the trajectory descends more or less steeply than 45°, is worse than arbitrary; its implied scenario diverges misleadingly from reality.

One point of figure 6.5, the superiority of making lift rather than drag for staying aloft—at least for Reynolds numbers high enough for decent lift-to-drag ratios—can be argued in another way. Consider a hypothetical drag-based descender that loses altitude at the same rate (0.41 meters per second) as the winged seed-leaf of the Javanese cucumber *Alsomitra* (*Zanonia*), which operates at $Re=4000$ (Azuma and Okuno 1987; Alexander 2002). If the descender weighed no more than that seed-leaf (210 milligrams) and took the form of a flat horizontal disk (thus normal to the upward relative flow: $C_d=1.2$), it would need an area of 3.4 times that of *Alsomitra*'s 0.005 square meters. And the latter operates at the unimpressive lift-to-drag ratio of 3.7, apparently accepting a lesser value than its best 4.6 to gain the intrinsic stability critical for a totally passive glider. For a similar reason, about a thousand years ago, (lift-based) windmills whose blades rotated in a plane normal to flow became common and displaced (drag-based) ones that turned horizontally like cup anemometers. Ships with propellers displaced most drag-based side-wheel and stern-wheel boats, starting a century and a half ago. And both transitions happened long before the aircraft-stimulated development of propellers that could achieve particularly good L/D ratios.

The dichotomy between drag-based and lift-based descent-slowing carries a further message. That size-dependent shift from drag as good to drag as evil may constitute an odd adaptive barrier—a device well-attuned to one mode will ordinarily be especially bad in the other. Active flight has evolved from gliding flight whenever it has appeared, but gliding flight seems never to have evolved from drag-based descent retardation. A suggestion that flying insects took that route (Wigglesworth 1963) has never gained substantial support. The nearest thing to an evolutionary switch I can think of occurs in a few Lepidoptera such as the gypsy moth (*Lymantria dispar*), a notorious pest in North America. Instead of basing dispersal on actively flying adults, the first instar caterpillar does the job by paying out long silk strands as if a newly hatched spider. So one animal does both, but when very different in both shape and size.

Not that one can't imagine plausible designs that might permit fairly easy shifts from, say, drag maximization to gliding, at least in passive craft. A round horizontal disk with a mass on a rigid stalk beneath its center will descend with lots of drag. Moving the stalk and mass closer to an edge could convert the device to something like a hang-glider, with better still-air dispersal distance as a selective reward. I would not place a bet, even at good odds, against the reality of such a scenario—some seed-leaves look like good candidates.

FLYING—WHY BIG CRAFT SHOULD FLY SWIFTLY

In simple gliding, gravity provides the motive force, and energy to sustain the process comes from the progressive loss of altitude; in soaring, the energy ultimately comes from atmospheric structures. In the sustained, active flight of airplanes, birds, pterosaurs, insects, and bats, the lift of paired wings again plays the key role. But sustaining altitude without that gravitational or atmospheric free ride demands some engine, typically either a propeller directing air rearward with a fixed wing deflecting the craft's airstream downward or else flapping wings that create both rearward and downward airstream momentum. Averaged over all but the briefest of time spans, the upward aerodynamic resultant must precisely equal the downward gravitational force, the weight of the craft, just as in steady gliding. So the same basic scaling rule appears applicable. As in equation (6.10), weight divided by wing area, or wing loading, ought to vary directly with body length for an isomorphic set of fliers or, assuming constant density as well, with body mass to the one-third power. And similarly, bigger should mean faster; from equation (6.11) we see that

$$v \propto (W/S)^{1/2} \propto l^{1/2} \propto m_b^{1/6}, \tag{6.13}$$

which is a specific prescription for how much faster larger aircraft must fly.

In a lovely book, Tennekes (1996) makes this a major point, drawing a single line on a graph that appears to indicate compliance (without even a shift in the constant of proportionality, 0.38 in SI units) from fruit flies to the largest passenger aircraft at the time, a Boeing 747. Wing loading, W/S, goes up as the cube root of mass, m, and equation (6.13) predicts cruising speed quite well. Other sources such as McMahon and Bonner (1983), Azuma (1992), and Dudley (2000) cite the same rule. Airplanes fit almost perfectly, at least if one excludes gliders and human-powered craft, which keep wing loading and therefore cruising speed deliberately low. Birds fit the same regression line, wing loading again going up with mass$^{1/3}$, and with the same proportionality constant.

Insects, though, scatter a lot more, with the scaling line recognizable only by lumping some very lightly wing-loaded butterflies and moths with heavily loaded beetles, bees, and flies and by following downward the pre-established trend. Except for dragonflies, the insects we regard as smooth, fast fliers weigh several times more relative to their wing areas than the scaling relationship predicts—as do hummingbirds. Furthermore, equation (6.13) predicts flying speeds considerably in excess (roughly double) what the all-too-few reliable measurements (and other considerations) indicate.

Why this fly-in-the-ointment? I think nothing especially obscure underlies the deviation. In a sense, the problem combines etymology and entomology (my apologies to the hummingbirds). The smaller the flapping flyer, the more the function of what we call a wing approaches that of a propeller and the less it resembles that of a paradigmatic airplane wing. Indeed, the greatest deviations from the rule occur where wingbeat frequencies are highest. More specifically, they occur where the speeds of the wings in their upstroke-downstroke oscillation most exceed those of the animal's forward flight. What's going on is that a flapper uses its wings more often than does a fixed-wing craft. For that reason, the speed most relevant becomes the tip speed of each beating wing rather than the forward speed of the craft. (Alternatively, the area most relevant becomes the area swept by the wings in a stroke rather than the area of the wings themselves—"disk loading" thus replaces wing loading.)

So we need another parameter, the ratio of the forward movement of the craft to up-and-down wing movement. The propeller designers provide one, the so-called advance ratio, J, although for applicability to animal flight it has to be altered slightly. As a wing swings down and up, it goes through less than an angle of 360°, and the resulting additional parameter, amplitude, can itself vary. As usually given for flying animals (Ellington 1984),

$$J = \frac{v_f}{2\phi fR},$$

(6.14)

where v_f is flight speed, ϕ amplitude (or "stroke angle"), f wingbeat frequency, and R wing length. (In practice, amplitude varies too little to matter much, but the formula properly recognizes its basic relevance.)

Small insects suffer more from the pernicious effects of viscosity and must make do with lower L/D ratios, as we saw earlier. So they have to beat their wings at high frequencies and fly slowly—their wings go up and down a lot for only a little forward progress. J for a bumblebee peaks at about 0.66, for a black fly 0.50, for a fruit fly 0.33. By comparison, ducks and pigeons fly at about 1.0 (Vogel 1994b). Halving the advance ratio roughly doubles the effective wing area since the wings get used more often. That's consistent with what we see when comparing birds, which follow the scaling rule, with the faster insects, which have greater wing loading and fly faster than the rule predicts. For the particulars of how small insects achieve frequencies that may reach 1000 per second—fabulously lightweight wings, special neuromuscular devices, and so forth—one should look at Dudley (2000) and other specialized accounts.

From the viewpoint of scaling, the relatively high flight speeds of some tiny insects—around 1 meter per second for a fruit fly (500 body lengths per second; a duck does less than 100)—might be surprising, whatever their obvious utility in an atmosphere that is rarely still. After all, their higher surface-to-volume ratios imply a relatively greater cost of dealing with body drag and a lesser cost of offsetting gravity. In fact, while true, the force needed to oppose drag remains modest—for a falcon less than 1 percent of weight or lift (Tucker 2000); for a teal (a duck) about 2 percent (Pennycuick et al. 1996); for a desert locust, about 4 percent (Weis-Fogh 1956); for a bumblebee, 8 percent (Dudley and Ellington 1990); for a fruit fly still only about 10 percent (Vogel 1966). Those percentages, incidentally, suggest that drag reduction through streamlining can only marginally affect the cost and practicality of flight. Gravity provides the more potent enemy.

Both whole organisms and their parts inevitably exceed the density of air—nature makes no blimps or ascending balloons. So any biological system that keeps a bit of atmosphere between itself and the earth's surface must contend with gravity. The particulars prove complex physically, complex biologically, complexly size-dependent, and at least occasionally counterintuitive.

One final example should persuade the reader of that last assertion. We expect that a greater gravitational acceleration, as would characterize a larger planet, would make passively aerial organisms descend faster. That same increase in g, though, ought to increase atmospheric density—in one scenario (which we will assume) increasing directly with g (see, for instance, Taylor 2005). In the Stokes' law world of the small and slow, terminal descent speeds will indeed increase. They depend on the difference between the densities of object and medium, and even doubling the latter will still leave it insignificant. The change in atmospheric density will not affect the viscosity, on which drag depends; thus weight goes up, drag does not.

At Reynolds numbers above one, drag becomes increasingly dependent on atmospheric density (eq. 6.12) and decreasingly dependent on viscosity. Ignoring some complications, drag will vary directly with density. If both drag and weight vary directly with gravitational acceleration, then drag-based terminal descent speeds will not—which strikes one as odd. By contrast, if lift also varies with density (eq. 6.11), then lift-to-drag ratios and glide angles (eq. 6.9) will be independent of air density. So the increased weight of a glider will make it descend faster—as in the drag-based Stokes' law range but not the drag-based higher Re range. Yes, gravity always drives the aerial system earthward, but how descent speed scales with changes in g varies rather peculiarly with both size and mode of travel.

CHAPTER 7

Gravity and Life on the Ground

Nuisance, Necessity, Non-Event

Unless actively counteracted, gravity inevitably makes aerial life descend. For terrestrial life, gravity's roles are less obvious, less immediate, less consistent. Sometimes it matters; sometimes other agencies eclipse its effects. Sometimes it acts as impediment or nuisance; sometimes it plays a crucial positive role. While not exerting quite the same physical dominance, gravity has more diverse consequences and has elicited a wider range of biological devices for organisms that live out their lives on the ground.

For one thing, much more depends on the distinction, made in the last chapter, between gravity, thus weight, and inertia, thus mass. Steadily lift an object, and you work against gravity; pull downward, and you enlist gravity's assistance. Sliding an object steadily sideways may entail no irreducible resistance, but the frictional force you do feel still comes from its weight, from the pressing of the object against the substratum. But accelerate an object, and you work against its mass. Big Neanderthal thrusting spears put gravity to use, working best with heavy bodies that leaned forward over well-planted feet to get sufficient purchase on the ground. Lighter, thrown spears depended more on inertial mass—a running body, in effect no purchase at all, could aid a launch. Similarly, a lighter person has to lean farther outward when opening a substantial door. That more effectively applies the person's lower body to produce the sideways force that accelerates the mass of the door. Muscularity matters little.

For another, organisms consist of both solids and liquids. In practice the two phases of matter face gravity in slightly different guises, repercussions of the difference between compressive stress and hydrostatic pressure. Both variables have dimensions of force per unit area, but in a specific direction for stress while omnidirectional for pressure. Stack solid bricks ever higher (with pads between to ensure uniform force transfer) and eventually the lowest will crush. That crushing point is reached when the compressive strength of brick, or force-resistance relative to cross-section, of about 20 megapascals (or meganewtons per square meter) equals the weight of the column relative to cross-section. If made of bricks whose density is 2000 kilograms per cubic meter, the column will be about 1000 meters high. Taper changes the picture—a column tapering upward can extend farther, since the greater weight near the bottom will be borne over a greater cross-section. Conversely, one expanding upward will crush its bottom at a lower height.

Extend a pipe of liquid water upward in the air, and the pipe eventually bursts at ground level. The column of water extending upward stresses (in the sense above) the material of the pipe, but it does so in proportion only to the height of the column—neither cross-section, taper, or contained volume bear on the situation. The pressure difference, Δp, across the walls of the pipe will be the product of the liquid's density, ρ, gravitational acceleration, g, and the column's height, h, in the familiar equation for both manometry and conversions of pressure units:

$$\Delta p = \rho g h \qquad (7.1)$$

Transforming that pressure to tensile stress (σ_t) in the wall of the pipe depends, obviously, on the thickness of the wall of the pipe (Δr, assumed well below r itself) and, less obviously, on its size, here the radius:

$$\sigma_t = \frac{\Delta p\, r}{\Delta r} = \frac{\rho g h r}{\Delta r} \qquad (7.2)$$

This last equation is pregnant with biological implications. For a given pressure and a wall material of a given tensile strength, a narrower pipe (lower r) will manage with a thinner wall (Δr). For example, your capillaries withstand pressures about a third as high as that in your aorta despite having walls two thousand times thinner. They manage that apparently paradoxical feat (convenient for material exchange) because their diameters are about four thousand times less than that of the aorta. As one can see from equation (7.2), they feel about six times less tensile stress in their walls rather than the many times more that one might guess (Zweifach 1974; Caro et al. 1978). Or, anticipating just a bit, since neither cardiac blood pressure (Δp) nor maximum muscle stress (σ_t) changes with body size, then the thickness of the ventricular wall (Δr) can remain a constant fraction of heart radius (r) itself (Seymour and Blaylock 2000).

Here I'll examine three situations in which gravity plays a role, asking what sets blood pressures for animals of different sizes and with what consequences; what determines the gait transition speeds for legged animals; and what sets the heights of trees and forests.

CIRCULATION AND HYDROSTATICS

The scaling of the circulatory components of vertebrates, especially mammals, has recently enjoyed a renaissance of attention. Heart mass and total blood volume increase in direct proportionality to body mass ($\propto m_b^{+1}$). Capillary length goes up slightly with mass ($\propto m_b^{+1/5}$), while capillary density ($\propto m_b^{-1/6}$) and maximum heart rate ($\propto m_b^{-1/5}$) go down slightly. Maximum oxygen consumption and cardiac output go up but not as fast as body mass itself (both $\propto m_b^{+7/8}$). But not all variables vary with mass. In particular, blood viscosity, capillary and red blood cell diameters, aortic flow speed, and average arterial blood pressure remain nearly the same. (Exponents from Baudinette 1978; Calder 1984; Dawson 2005.)

In looking for gravity's consequences, we might take a closer look at that size-independence of blood pressure. That constancy, first noted over half a century ago, has become ever better supported. Recognizing that pressure cycles non-sinusoidally between systolic peaks and diastolic minima, an effective average is commonly taken as a third of systolic plus two-thirds of diastolic. For mammals, that average is about 12,900 pascals (97 millimeters of mercury). So we humans are typical, with systolic pressure of about 16,000 pascals (120 millimeters of mercury) and diastolic pressure of 10,500 pascals (80 millimeters of mercury) and an average of 12,300 pascals (93 millimeters of mercury). For birds average pressure runs somewhat higher, 17,700 pascals (133 millimeters of mercury) (Grubb 1983). The "lower" poikilothermic vertebrates develop lower pressures.

From our present viewpoint, constancy of blood pressure seems paradoxical. Terrestrial animals amount to ambulatory manometers, obeying equation (7.1), with a blood density of about 1050 kilograms per cubic meter for ρ, and thus with a pressure gradient of 10,300 pascals per meter from head (lower) to toe (higher). Without auxiliary pumps, blood pressure at head height has to drop as body height increases. Thus a normal human has a diastolic blood pressure of about 5300 pascals (40 millimeters of mercury) in the head and 20,000 pascals (150 millimeters of mercury) in the feet (Schmidt-Nielsen 1997). Health-care professionals learn to cuff the arm at heart height when taking blood pressures, although (by my informal survey) almost none of them know just why they're told to do it or what error an improper cuff height introduces.

While gravity cannot be turned off, the relatively high pressure gradient needed to keep blood flowing through the resistive vessels ordinarily exceeds that gravitational gradient. We fix that problem with passive one-way valves in the veins of our extremities to persuade skeletal muscle to help with pumping. The fix ordinarily works, although a human maintained upright without working the muscles of the legs soon loses consciousness. So with that 5300 pascals (a little lower if hypotensive) and a bit of help we manage to keep blood flowing steadily and to keep our brains decently supplied with oxygen—I've seen no claim that mental agility decreases with body height.

An animal with its head a meter above its heart should be in serious trouble at standard mammalian cardiac output pressure. Its effective average pressure will be only 2600 pascals (20 millimeters of mercury), far below the normal 12,300 pascals. Half a meter should be about the limit, with gravity dropping diastolic pressure by 5100 pascals (almost 40 millimeters of mercury).

As it happens, animals that hold their heads high do not have normal mammalian blood pressures. Most sources of scaling exponents include some parenthetical remark such as "excluding the giraffe" (Calder 1984) after noting the standard and its near size-independence. That exclusion represents not some special case but a necessary threshold for gravitational compensation. We might view the situation with the aid of a dimensionless ratio, a "gravitational hazard index" (GHI). Such an index expresses the height of an animal in pressure units, that is, as if it were a blood-filled manometer obeying equation (7.1); it divides this "manometric height" ($\rho g h$) by average arterial (heart-high) blood pressure:

$$GHI = \frac{\rho g h}{\Delta \bar{p}}. \tag{7.3}$$

Figure 7.1 considers average mammalian blood pressures relative to this GHI. (One should bear in mind the great lability of one's own pressure and recognize the limitations of data from animals of even less assured disposition.) Blood pressure seems to have not one but two limits. A lower horizontal line must represent the minimum average arterial pressure needed to overcome the resistance of the systemic system of conduits. It has a value of about 10,000 pascals (75 millimeters of mercury). A vertical line to the right of the data points represents the limit set by the need to supply a brain at some minimum pressure after gravity takes its toll of cardiac output. It appears to have a value (dimensionless) of about GHI=1.7. Of course the numbers have some systematic bias from using overall height instead of heart-to-head height and zero rather than a figure for minimum tolerable cranial pressure.

Small mammals can ignore gravity, while large (at least tall) ones most definitely must care. In effect, mammals tolerate the gradual diminution of cranial blood pressure with increasing height—up to a point. That point corresponds to animals only slightly taller than ourselves. (And thus slightly hypertensive or unusually tall humans manage quite well.) While

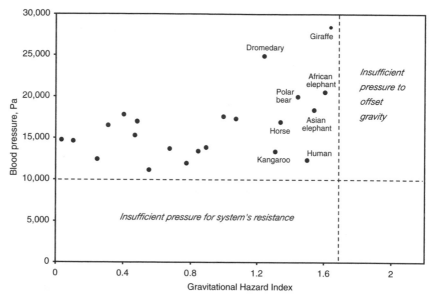

Figure 7.1. Blood pressures of mammals plotted against gravitational hazard index (*GHI*). Most of the pressure data come from Seymour and Blaylock (2000), with a few additions and confirmations from www.ivis.org; heights have been estimated from the photographs and shoulder heights in Nowak (1991).

we have too few reliable data for mammals taller than ourselves, we have no reason to suppose that their blood pressure does anything other than track the sum of two components, that set by the resistance of the system and that set by the need to raise blood against gravity.

The perceptive reader may think of a simple evasion of this problem of getting blood up to the head—in a word, siphoning. Since vessels are full, descending blood could draw blood upward, reinvesting the energy of descent to raise blood. The issue of siphoning has provoked no small amount of controversy; at present the weight (sorry!) of evidence opposes it (Pedley et al. 1996). Tall mammals do generate the high pressures needed to raise blood without siphoning—one of the incentives for work on the physiology of giraffes, providing circumstantial evidence. In addition, blood vessels in the head appear reinforced against conventional outward aneurysms rather than inward collapse. And the descending veins are all too collapsible, so blood commonly descends in boluses rather than a continuous stream. Of course one shouldn't rule out the possibility that some siphoning might occur under some circumstances, perhaps during vigorous aerobic activity.

At the same time, blood vessels in the legs must be strong enough to take both the extra cardiac pressure and the extra gravitational component. Which, not surprisingly, they are. Beyond that the entire legs must be wrapped with an especially inextensible integument lest the extracellular space become oedematous. Which they are as well. In giraffes in particular the vessels of head and neck have similar reinforcement, almost certainly important for preventing aneurysms when an animal lowers its head to drink.

Between their higher average blood pressures and the absence of very tall ones, contemporary birds should never hit an equivalent limit. One does wonder about the giant moas of New Zealand, extinct for the past four hundred years—the wall thickness of some miraculously preserved artery would probably allow estimation of their blood pressure. Reptiles, by contrast (or "other reptiles" to some) present a much more interesting issue. Blood pressures run about a third of those of mammals, so the inflection point of figure 7.1 should occur at a third of the equivalent mammalian body height—about 0.57 meters rather than 1.7 meters. Most extant reptiles are either small or lie low to the ground, so they should have no problem with gravitational pressure drop even with their low pressures. Not all, though; in particular, some fairly long snakes climb trees and surmount obstacles, making "fairly long" into "fairly tall." Unsurprisingly the average heart-level blood pressures of long snakes vary widely, from about 3300 pascals (25 millimeters of mercury) in aquatic species to around 10,500 pascals (80 millimeters of mercury) in terrestrial climbers. More remarkably, terrestrial climbers position their hearts substantially closer to their anterior ends—in a comparison of a python and a file snake of about equal length about 25 percent versus 37 percent of snout-vent length. In addition to these differences, climbers have reinforced body walls ("g-suits") around their posterior regions and especially well-developed baroregulatory reflexes (Seymour and Arndt 2004).

No basic inferiority of reptilian heart muscle should rule out the giraffe's trick. More likely, the main trouble lies in their basic lung-shunting scheme, dividing cardiac output between interconnected systemic and pulmonary circulations. We mammals (and birds) have no such connection and an unalterably serial circulation. Volume flow (Q) through the lungs must exactly equal volume flow through the systemic circulation, depriving us of the ability to reduce pulmonary flow during, for instance, prolonged or deep diving. But we gain the ability to run the pulmonary circuit at a different pressure (Δp) (typically a fifth or sixth) than that elsewhere. In effect, we keep the cost (ΔpQ) of pulmonary pumping low with a reduced Δp; reptiles keep the cost low with a reduced Q. (Crocodilian reptiles, with optional shunting, may have the best of both worlds; but the present problem is moot since they live in a severely horizontal or aquatic world.)

Most extant reptiles may keep their heads down, but one wonders about dinosaurs, some famously tall. To take an extreme case, *Brachiosaurus* may have carried its head as much as 8 meters above the heart, with an overall height of 12 meters (Gunga et al. 1995). A *GHI* limit of 1.7 suggests an average heart-level blood pressure of 73,000 pascals (550 millimeters of mercury). Recognizing their atypically low hearts and using $(10,000 + \rho g h)$ instead raises the estimate to 92,000 pascals (690 millimeters of mercury). Either figure far exceeds the arterial pressure of a giraffe. One supposes that *Brachiosaurus* kept its head up—as Carrier et al. (2001) point out, a lowered head would have extended so far forward from the center of gravity so as to have severely impeded turning. And their vertebrae certainly permit high-headedness. Still, we can imagine several solutions or evasions. They may have tolerated brief cranial anoxia. Or perhaps dinosaurs had subambient cranial blood pressures, driving flow by the pull of siphons rather than the push of pumps. A partial solution may not be especially obscure. Birds evolved from (or are) dinosaurs, and birds have fully serial circulatory systems like ours. That dinosaurs had essentially avian systems thus involves no great stretch of any evolutionary scenario, according to one of their intimates, Kevin Padian (personal communication).

To Walk or to Run

Almost all our terrestrial vehicles move on rotating wheels. Occasionally we even use temporary, axle-less wheels, moving heavy objects on rollers by shifting them from rear to front as rollers emerge, one by one. Physics imposes no irreducible minimum cost on terrestrial locomotion if it's on the level. Only imperfect stiffness of wheels and path, friction of wheel bearings, accelerations, slopes, and air resistance impede motion. Railroads, with metal wheels and level, metallic tracks, could provide economic transport with the inefficient steam engines of two centuries ago, long before road vehicles could shift from draft animals. Wheels, especially with axles, are splendid devices.

No terrestrial animal goes from place to place on wheels and axles. One can argue (as did Gould 1981) that evolutionary constraints preclude their appearance. Or one can argue (as did LaBarbera 1983) that we easily overrate the utility of wheels, that they lack versatility and, in particular, that they work badly on either soft or bumpy surfaces. That latter argument receives at least tacit endorsement by recent attention (mainly military) to legged robots for off-road use, emulating the general arrangements of animals such as ourselves.

Legged transport may be widespread, but it can't be described as energetically efficient. However many legs an animal uses, it faces a basic difficulty that rolling wheels circumvent. Legs work by reciprocating rather than rotating, which means that any leg of finite mass must waste work accelerating at the start of a half-cycle and then decelerating again at the end. Of course an evasion comes immediately to mind—bank the decelerative work for reuse in the subsequent acceleration. What kind of short-term battery, then, might store that work? Electrochemical storage could be used, like the regenerative brakes of some hybrid automobiles, but no natural examples have yet come to light. Or inertial storage might serve, as in a flywheel. Again we can point to no obvious natural case, although bicycles, remarkably efficient locomotory prostheses, make some use of the scheme.

Two kinds of brief batteries do find widespread use—lifting and then lowering masses against and with gravity, and straining and then releasing springs. Interestingly, animals cannot be dichotomized by their use of one or the other of these fundamentally different ways to store energy. Instead, legged terrestrial animals typically depend on both, shifting from one to the other at a specific speed. At low speeds, gravitational energy storage does the job in what we call walking gaits; at higher speeds elastic energy storage serves in the various running gaits. It would be a rare culture that lacks specific words for at least these two gaits, so obvious is the distinction.

The recognition that this shift in energy storage mode underlies the abrupt transition is recent—surprisingly recent. The traditional distinction notes that walking gaits have no fully aerial phase while running gaits include at least a brief aerial phase. True enough, except for elephants (at least), which trot without an entirely aerial phase, but it doesn't get us all that far. The realizations both that the basic game consisted of offsetting the inefficiency of legged locomotion and of the role of gait shifting in helping efficiency we owe to R. McNeill Alexander and his associates (Alexander 1976; Alexander and Jayes 1983, and other papers and books). In addition they have done as much perhaps as everyone else put together in working out its implications. The crux of the matter takes few words. In walking gaits, whether bipedal or polypedal, gravitational storage does the job, and almost the entire body mass contributes to the functional weight. In running and hopping gaits (trotting, galloping, cantering, skipping, bounding, etc.), elastic storage takes over, with stretched tendon storing most of the energy. Muscle and bone make only minor contributions.

How gravitational energy storage eases a task can easily be demonstrated. Sit on the edge of a desk, swing a lower leg back and forth, and measure its period. Plug that time, t, into the standard equation for a pendulum,

$$t = 2\pi \sqrt{\frac{l}{g}}, \tag{7.4}$$

and you get an effective length, l. My 1.1-second swing predicts a length of 30 centimeters, a reasonable measure of the distance from knee to the leg's center of gravity. The exercise isn't entirely trivial—it illustrates the ease with which one's neuromuscular system phases its output to maintain an efficient frequency. Try to change swinging frequency and you'll find yourself working a lot harder. Put on a heavy shoe and you'll swing with a longer period, again with little or no initial awkwardness. Similarly, when walking, you immediately adopt a "natural" pace, increasing or decreasing speed as much by changing stride length as by changing frequency. A pendulum length for a normal adult pace of 1.4 seconds per stride is about 50 centimeters. That's not unreasonable for hip to center of gravity of a leg—ignoring some bias and complications from the constrained motion of a leg in contact with the ground.

About the location of the pendulum, though, extrapolating from leg swinging to walking misdirects us. Just how gravitational storage operates in walking gaits turns out to be less easily specified. Indeed, it works in quite an odd way—perhaps the reason it so long escaped analysis. Were our walking to resemble the swinging of an ordinary pendulum, we would reach greatest speed and our centers of gravity would be lowest in mid-step. In fact, we are highest, not lowest, and slowest, not fastest, in mid-step, as we vault over relatively extended legs. In addition we sway slightly side to side at half the frequency at which we move up and down.

Walking, again whether bipedal or polypedal, is commonly described in terms of the motion of an inverted pendulum. The head and torso provide almost all the relevant mass whose center of gravity matters, rather than the mass of the legs, despite their more rapid motion. As shown in figure 7.2, head and torso travel in a series of arcs, convex upward rather than downward, as in a conventional pendulum. The way an egg rolls end-over-end down a slope gives some idea of the way kinetic and gravitational energies interchange—speed and height of center of gravity peak at opposite phases of its motion. While an inverted pendulum doesn't provide an intuitively obvious physical model, the analogy turns out to be analytically powerful.

One might think of walking as a process of lifting one's center of mass and then allowing it to fall forward, the combination forming an arc.

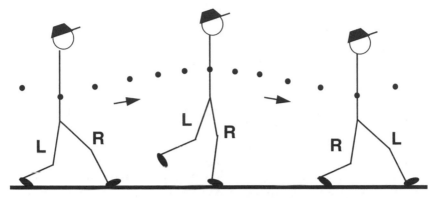

Figure 7.2. The motions of the body in a half-step of walking. At mid-stride the body is highest and the speed (indicated by distances between spots at standard intervals) is lowest. Changes in both have been exaggerated. (Adapted from Vogel 2003; see also Biewener 2003.)

Gravity sets the downward acceleration of that forward fall which in turn sets a practical limit on stride frequency. That insight let Alexander and Jayes (1978) estimate maximum speeds of walking with only a few empirically supported assumptions. First, in walking, at least one leg must always be on the ground—that is, the "duty factor" or temporal ground contact fraction cannot be less than 0.5. And second, relative stride length—stride length over hip-to-ground length—should peak at the same value for walkers of any size. Finally, the walkers should be similarly proportioned and walk with similar maximum arc lengths for their strides. They predicted that the limit on downward acceleration would limit walking speeds to no more than about 0.4 or 0.5 times a particular dimensionless ratio, v^2/gh. The latter is the quotient of forward speed, v, squared to gravity times a height, h, this last taken as that of the hip joint from the ground.

The ratio happens to have the same arrangement of variables as that between kinetic energy and gravitational potential energy,

$$\frac{mv^2}{mgh} = \frac{v^2}{gh},\qquad(7.5a)$$

where m is body mass and the factor of 2 in kinetic energy has been ignored. It also appears if inertial force is divided by gravitational force,

$$\frac{ma}{mg} = \frac{v^2}{gh} = Fr,\qquad(7.5b)$$

noted as the Froude number, *Fr*, in the last chapter—the ratio introduced by William Froude in the nineteenth century as a scaling rule for models of the hulls of ships. In this last guise, it will reappear in the next chapter. It can also be obtained by ignoring the constant factor and squaring what's left of both sides of equation (7.4), which itself can be obtained by simple dimensional analysis.

The ratio provides a specific rule for the way maximum walking speed varies with animal size, a rule with both explanatory and predictive value. The rule applies to a very wide range of walkers, all of which hit maximum speeds at Froude numbers between 0.3 and 0.5 (Biewener 2003)*. Above that range of size-adjusted dimensionless speeds, animals switch to other gaits—we begin to jog, a dog begins to trot, a crow begins to hop. The greatest distance covered per unit energy expenditure occurs at about *Fr* = 0.25, the size-independent optimum walking speed. The diversity of organisms that follow the rule makes it a remarkable generalization. It's about the best illustration of how dimensionless ratios can serve biomechanics just as they serve mechanical (mostly fluids) engineering. Had Alexander not pointed out Froude's precedence (albeit in relating at the lengths and speeds of surface waves), we would now be talking about the Alexander number. Beyond that, Alexander's argument about minimizing cost of transport has recently received strong theoretical support from recent work by Srinivasan and Ruina (2006) on models moving with both real and hypothetical gaits.

Animals of whatever size subject their bones to about the same maximum stress when moving—about twice standing during walking and about five times standing in running—but not exceeding 50–100 megapascals (Biewener 1990). Using this range of maximal bone stress and the range of transition Froude numbers one can ask about the speeds of dinosaurs. The combination implies that the largest theropods such as *Tyrannosaurus* ran gingerly if at all (Alexander 1976; Hutchinson and Garcia 2002); conversely, they could walk exceedingly fast. And from the skeletal dimensions and trackways the walking speed of the 3-million-year-old hominids of Laetoli, Tanzania can be estimated. They were about a third shorter than modern humans and should have been slower by a similar factor (Alexander 1984).

We can also ask what might happen were the value of gravitational acceleration altered. Greater *g* should give a higher transition speed; lower *g* should give a lower transition speed. Humans on the moon, with a sixth of terrestrial *g*, found that hopping was a better way to get around than walking, which would have been (ignoring the effect of space suits) less

---

*Sometimes the square root of equation (7.5b) goes as the Froude number; if so, the transition values are 0.55 to 0.7

than half as fast as on earth. Skipping, as done here on earth by children, was a practical gait as well (Minetti 2001). When walking on a (terrestrial) treadmill, partly supported by a traveling overhead harness, humans maintained the characteristic exchange of kinetic and potential energy of walking, but this simulated reduced gravity did demand an unusually high metabolic expenditure (Griffin et al. 1999); brief exposures in a maneuvering aircraft to truly altered gravity (Cavagna et al. 2000) gave similar results.

A good rule directs attention to apparent exceptions. Emperor penguins walk long distances at an especially high cost for their size. Their short legs make them geometrically dissimilar to other birds—for their size, they make especially quick strides. While that may preclude the usual arrangement for energy interchange, they have another, side-to-side waddling. The high cost, then, does not come from abandonment of the interchange, but from the high rates at which the muscles running their short legs must generate force (Griffin and Kram 2000). Penguin walking appears to be close to a model developed by Coleman and Ruina (1998), a bipedal toy or robot (a "passive-dynamic walker") that goes down a slope with a side-to-side pendulum motion—an easily built model can be found at http://ruina.tam.cornell.edu/research/topics/locomotion_and_robotics/. Incidentally, more humanoid models have now been developed; Collins et al. (2005) describe several.

Don't forget that on a level path, the entire cost of locomotion (ignoring drag) ultimately represents inefficiency. Although walking takes energy, the relative (mass-specific) cost of body transport decreases as the size of the animal increases. That most likely comes from a basic disability of muscle, the need to expend energy to produce force, even when moving nothing. The more rapidly we ask a muscle to develop force, the greater the cost, as just mentioned for penguins. The smaller the animal, the greater its stride frequency, and thus the greater the cost of level walking relative to its mass.

If the path slopes upward, walking incurs an additional cost, that of working against gravity, which scales with body mass. The combination of the cost of level walking with the additional price of going upward underlies a curious but familiar phenomenon. The relative difficulty of ascent depends on an animal's size. A horse walks more efficiently on the level than does a dog, but even a slight slope extracts a great fractional increase in demand for energy—quite familiar where and when animal-drawn vehicles transport. A small rodent handles slopes more easily than any dog, and those ants that construct roadways do so with magnificent indifference to slope, caring only about overall path length. Minetti (1995) applied treadmill data to predict the optimum slope of mountain paths, assuming a goal of gaining altitude cheaply. The slopes of paths in the Italian Alps corresponded nicely to his predictions, with switchbacks wherever the critical steepness would otherwise be exceeded. In theory, at least, one could predict the size of an unknown animal (perhaps a yeti) from the slopes of its paths.

To Trot or to Gallop

We bipeds have only a few variants on walking, such as flexed-legged rather than stiff-legged walking, race-walking, and goose-stepping. To these we add several gaits that depend on elastic energy storage, such as running, hopping, and skipping. Quadrupeds have a considerably wider range of gaits that use elastic storage; of these the two most common are trotting and galloping. In trotting, each of four legs strikes the ground in a left-right symmetrical sequence—front-left, hind-right, front-right, hind-left. In galloping, almost paired front and almost paired hind legs alternate, "almost" because one side usually leads. The minor asymmetry gives, for instance, the gallop of a horse its characteristic sound (Rossini 1829). Like trotting, galloping mainly stores energy from stride to stride as stretched tendon.

Several questions. First, why gallop? Simply because by doing so an animal can go faster. Among other things, galloping permits recruitment of an additional mass of elastic in the back and elsewhere for energy storage (Alexander 1988b). Moreover, after rising as trotting speed increases, cost relative to distance drops again following the shift to a gallop. The speeds of this second gait transition raise a peculiar question. Among quadrupeds that gallop, the trot-to-gallop transition occurs within a fairly specific Froude number range, between 2 and 3 (Biewener 2003). Froude number, again, represents a ratio of inertial to gravitational force. In this second transition, both gaits involve elastic rather than gravitational energy storage. So why should Froude number matter?

Perhaps we need to reverse the argument for the first transition. What determined that shift was the upper practical speed for walking. Here, by contrast, what matters may not be an upper limit for trotting but a lower limit for galloping, a limit set by the maximum practical aerial period. Trotting has (elephants, again, excepted) only short periods when no foot makes contact with the ground, while galloping involves considerably longer aerial phases. While airborne, an animal must fall earthward—with gravitational acceleration. Too long a fall, and an animal will not be easily able to position one or more feet on the ground beneath its torso. What can we make of that intuitive argument?

Assume an animal can allow itself to fall no more than a fixed fraction of its leg length, so

$$d \propto h \propto gt^2, \qquad (7.6)$$

where d is distance fallen, h is leg length, and t is the time in free fall. What we need to know is how the speed at transition, v, varies with leg length. Heglund and Taylor (1988) report that it varies as one might expect, with leg length divided by stride time—all gallopers basically gallop about the same way at the transition point. So

$$v \propto h/t. \qquad (7.7)$$

Combining the two proportionalities to eliminate t and taking the reciprocal (if you are constant, so is your reciprocal) yields the Froude number:

$$Fr = \frac{v^2}{gh}. \qquad (7.8)$$

Can we go a step further and rationalize the particular value (or range) of Froude number at which transition occurs? We might assume the empirical value and estimate the fraction of leg length that a galloper drops while airborne. Breaking speed into length per stride (l) and time per stride (t), we get

$$Fr = \frac{l^2}{hgt^2}. \qquad (7.9)$$

Heglund et al. (1974) reported a minimum galloping speed for a particular horse of 5.6 meters per second at a frequency of 2.0 hertz. Alexander et al. (1980) found that the stride length of a galloping horse is about five times its hip height. Adjusting that down from average to minimum speed (using the speeds of Heglund et al. 1974 and Heglund and Taylor 1988) gives a stride length of 3.4 times hip height, the latter about 1 meter (from a skeleton in our departmental collection). The final item needed is the fractional duration of the airborne periods at minimum galloping speed. Here specific data seem lacking—never mind our fanatic concern for maximum speeds. I'll assume two periods, each of 25 percent of stride duration, noting that relative time airborne will be at its lowest at minimum galloping speed.

These numbers give a stride duration of 0.69 seconds and thus airborne periods of 0.172 seconds each. During each period, gravity will make the horse fall 0.145 meters, about 15 percent of the hip to ground distance. That does seem a practical maximum for getting feet positioned for the next stride, again noting the very rough character of the estimate.

The Height of Trees

Trees surely provide the paradigmatic examples of gravitationally responsive macroorganisms. With a tall supporting column each keeps a crown of photosynthetic structures elevated in the face of a gravitational force that prefers otherwise. It does so to win access to sunlight in competition with other trees—that height cannot bring it significantly nearer the sun. Each of the lineages in which tree-like organisms have evolved from shrubbier or herbaceous ancestors has been built with the same basic material, wood. Each tree or tree-like system extracts water from the substratum and lifts it to leaf level, mostly through evaporation at the top and consequent suction below. Despite considerable structural and developmental diversity among the lineages, their tallest members have achieved about the same maximum heights, roughly 25 to 100 meters (Niklas 1997b). Explaining such consistency tests our understanding of the biological consequences of gravity.

Perhaps the column of stacked bricks invoked at the start of this chapter might provide an instructive analogy. Wood has a compressive strength of about 50 megapascals and a density of about 500 kilograms per cubic meter—as a ratio, ten times better than brick, incidentally. A column (as in figure 7.3a) could extend 10 kilometers upward even with no taper, a hundred times the height of the tallest contemporary tree. Resistance to compressive crushing clearly imposes no limit.

But crushing mainly afflicts short, wide columns. Failure will more likely result from so-called Euler buckling, the sudden collapse that occurs when the middle of a column bows ever further outward (as in figure 7.3b). Elastic moduli rather than compressive strength now become the relevant material properties. Fresh wood runs around 5 gigapascals (Cannell and Morgan 1987); the compressive moduli run slightly lower than the tensile (Young's) moduli, but trees compensate for the difference with some tensile prestressing. Trunk thickness becomes relevant because buckling stretches one side and compresses the other. The standard equation for Euler buckling (see Vogel 2003 or standard handbooks for mechanical engineers) gives a height of well over 100 meters for a trunk diameter of 1 meter. This assumes that the tree does not taper and that its entire weight is concentrated at the top—an unrealistically harsh scenario. Offsetting (at least in part) those biases, trunks are assumed straight and their bases firmly fixed. Even admitting the simplifications, though, gravitational loading should impose no serious risk of buckling.

We might look at the tree in yet another way, a simplified version of Greenhill's (1881) classic analysis. Consider a brief lateral perturbation

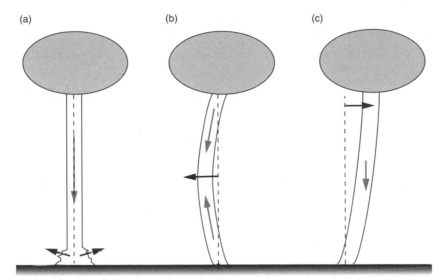

Figure 7.3. (a) A column failing by simple compressive failure—crushing. (b) A column failing by "Euler buckling," a mode in which, paradoxically, one side experiences tensile loading. (c) A column loaded in a manner similar to a cantilever beam, in which, once bent, its own weight generates a turning moment about the base.

near the top of a tree from wind or some other cause. That will move the center of gravity laterally, tending to make the tree topple. At the same time, it will generate an opposing elastic restoring force in the wood. In effect, this treats the tree as a self-loaded cantilever beam (as in figure 7.3c), albeit one extending upward rather than outward. If lever arm and restoring force scale linearly with deflection distance, then that distance drops out. Young's modulus also drops out since in practice it varies directly with the density of the wood. Again adopting standard equations, our 1-meter tree can extend upward about 120 meters before the wood of the tree reaches maximum tolerable stress. Again, trees rarely approach that value. And more realistic assumptions would raise the limiting height—I've once more assumed that the tree does not taper, which raises the center of gravity, and I've assumed that it pivots at the bottom rather than bending, which moves its mass too far outward.

Still, while both views—a column subject to Euler buckling and a cantilever beam—give unrealistically great heights, both say that height will scale with diameter$^{2/3}$. The girth of the taller tree will be disproportionately great, something easily observed. Quite a few sources mention that particular scaling rule, at least as far back as Greenhill's (1881) prediction.

In particular, McMahon (1973) cites it in his compilation of data on 576 trees in the United States, each of either record height or record girth for its species. The arguments for the rule have become much more sophisticated, in particular accounting for taper and crown weight. (Niklas 1992 has a good discussion.)

That each of several starting assumptions yields the same scaling rule gives us little help in choosing among them. Worse, we should not place great confidence in that exponent of 2/3, however often it gets cited. For few biological systems can we find so much data to test such a rule. For obvious reasons, books on practical silviculture, forest mensuration, and so forth pay great attention to the height and girth of trees. I tried a few regressions on published data and was rewarded with exponents ranging from about 1/3 to 4/3. Unsurprisingly, practical people concerned with timber production rely on more complex formulations—see, for instance, Johnson and Shifley (2002).

Moreover, many sources question the whole notion that the strength, density, and elastic moduli of wood determine the maximum heights and proportions of trees. The most common alternative views the limit as hydraulic, the problem of lifting water from the roots to such biologically prodigious heights. Our hearts develop systolic pressures during exercise of perhaps 25,000 pascals, and tall mammals when running probably approach twice that. Just working against gravity, a 100-meter tree has to move water against a pressure difference of 1,000,000 pascals, forty times better than our personal best. Worse, the main pump depends on suction from above rather than pushing from below, that is, on negative rather than positive pressure.

The main mechanism for raising water needs a few words, especially because at first encounter nearly every physical scientist expresses skepticism or outright incredulity. Evaporation across tiny interfaces in the feltwork of fibers of the cell walls of cells within leaves draws water out of the soil and up through a large number of small conduits (xylem) just beneath the bark. Surface tension at these interfaces (around 0.1 micrometers across) should have no trouble keeping air from being drawn in at the top—the surface tension of pure water across such a tiny interface can sustain a pressure difference of nearly 3,000,000 pascals, almost 30 atmospheres (Nobel 2005).

But then things get decidedly unconventional. Atmospheric pressure can push water up to a maximal height, defined by equation (7.1), corresponding to a pressure difference of 101,000 pascals at sea level—the presssure difference between one atmosphere and a full vacuum. For water (or xylem sap), with a density of 1000 kilograms per cubic meter, that height is 10.3 meters. Evacuate a vertical tube and place the open end in water, and the water will rise to that height, with a vacuum above. But if you fill a clean pipe a bit longer than 10.3 meters with water containing little dissolved gas, you may have to bully the system a little for the water level to drop and the vacuum to appear. In the interim the water column will have developed a pressure below zero pascals, a slight and brief but truly a negative pressure.

So even if water is freely available at ground level and can be raised without frictional losses, trees should be able to grow no higher than 10.3 meters—unless they capitalize to a fabulous degree on such negative pressure. Before taking offense at the notion of negative pressure, pause to observe the liquid, not gaseous, state of the water. The internal intermolecular cohesion that makes a liquid a liquid rather than a gas should render it perfectly capable of withstanding tensile stress, the respectable term for negative pressure. The tricky part is containing a liquid while subjecting it to tensile stress. Not only must its intermolecular cohesion withstand the stress, but so must its adhesion to the walls of the container— if either grip fails, a vacuum will appear. In addition, very little gas or other material can be dissolved in the water, so ordinary soil water must be rigorously preprocessed.

Trees apparently meet these demanding conditions and raise sap despite severely negative pressures. A field-usable device (a so-called Scholander bomb—see Scholander et al. 1965) makes possible routine (if indirect) measurements of negative pressures in plants by reading the positive pressures required to counterbalance them. Pressures of -1 or -2 megapascals (-10 or -20 atmospheres) are common, and values as extreme (one hesitates to say "as high") as -12 megapascals (-120 atmospheres) have been reported (Schlesinger et al. 1982). In laboratory tests, macroscopic quantities of water have resisted tensile stresses of hundreds of atmospheres, so the picture does not rely solely on calculated intermolecular forces. (But Abraham Stroock tells me that high values of relative humidity at the evaporative surface may impose an additional limit on maximum possible negative pressures.)

Other things being equal, the taller the tree, the more extreme the negative pressures. And the more extreme the pressures, the greater the danger that liquid within some conduit will cavitate, interrupting the process and putting that conduit out of action just as if it were an unprimed pump. Cavitation does occur with some regularity—this is no hypothetical hazard—sometimes embolizing a large fraction of the conduits in a normal tree. In practice, the greater the diameter of the conduits running up the tree, the greater the likelihood of cavitation (Ellmore and Ewers 1986; Maherali et al. 2006). But other work (see, for instance, Holbrook and Zwieniecki 1999 and other papers by each of these authors) has uncovered specific devices to minimize the propagation of embolisms and to repair embolized conduits.

Trees face a curious balancing act. Their demands for water vary over a wide range, low in conifers, for instance, and high in many broad-leaved trees. Beyond the gravitational loss of 9800 pascals per meter (from equation 7.1), the movement of sap causes an additional loss, that due to the fluid-mechanical resistance of the conduits. The general rule for pressure drop per unit length ($\Delta p/l$) due to laminar flow in circular conduits is the Hagen-Poiseuille equation, here given in terms both of total flow, Q, and of maximum (axial) flow speed, v_{max}:

$$\frac{\Delta p}{l} = \frac{8\mu Q}{\pi r^4} = \frac{4\mu v_{max}}{r^2}. \tag{7.10}$$

μ is the fluid's viscosity and r the radius of the conduit. Whether one considers total flow or flow speed, the smaller the conduit the worse the pressure drop. In addition, additional losses occur as sap passes between adjoining conduits—see, for instance, Lancashire and Ennos (2002). One might argue that a tree should move water in pipes large enough to keep the cost of flow low but not so large that embolizing becomes an excessive risk. Of course in enlarging pipes to reduce losses from flow, trees eventually meet diminishing returns—after all, that gravitational loss of 9800 pascals per meter remains.

Thus we expect conduit sizes will strike a balance, large enough to keep flow losses down to the same order as gravitational losses but not much larger. After all, why should a tree risk making conduits large enough to reduce flow loss much below the unavoidable gravitational pressure loss? What do we find? Maximum flow speeds in vivo can be measured by heating a trunk locally and then timing the interval before a thermocouple located somewhat higher detects a temperature change. I calculated pressure drops per unit length for a variety of trees (and a liana) from a variety of sources, using measured averages of maximum speeds and conduit diameters from Milburn (1979); Zimmermann (1983); Gartner (1995); and Nobel (2005). The data cover a ten-fold range of diameters and a hundred-fold range of speeds; the resulting pressure drops range from 1300 to 20,000 pascals per meter, that is, from 13 to 200 percent of the gravitational drop, with little evident regularity. These highly heterogeneous data reflect spread in conduit diameters within individual trees, uncertainty about which ones happen to be active and not embolized at a particular time, variation in flow speeds with time of day and wetness of season, and so forth.

Nonetheless, the values don't disagree with the expectation that trees balance the diminishing returns and increasing risk of enlarging conduits with a reasonable coincidence of flow and gravitational losses. At the same time, the values provide at least indirect support for the idea that the difficulty of lifting water imposes a general limitation on forest height.

Recently Koch et al. (2004) measured an extreme pressure of –1.8 mega-pascals 4 meters below the top (112 meters) of the tallest known tree, a redwood (*Sequoia sempervirens*). (That, perhaps coincidentally, is just under the –2 megapascals of equal gravitational and flow losses.) They found that in the laboratory a pressure of –1.9 megapascals imposes serious loss of hydraulic conductivity on such a tree and therefore argue that hydraulics limits height. But I wonder if the closeness of those figures, –1.8 and –1.9 megapascals, merely tells us that such redwoods conduct and utilize water no better than they have to, given their heights.

The argument also doesn't square with the way gravitational pressure losses and predicted flow may often be the lesser part of the overall negative pressures at tree-top heights. Major pressure drops come from extracting water from less-than-saturated soil ("matrix potential" sometimes), from osmotic processes in roots, and (as noted) from flow through the pits and plates that divide the ascending tubes of xylem. Trees 20 or 30 meters high often develop pressures of –2 megapascals or more, far above a twice gravitational drop of –0.4 to –0.6 megapascals. For that matter, the record of –12 megapascals mentioned earlier comes from measurements on a desert shrub, not a tree, and mainly results from the difficulty of sucking liquid water out of very dry soil.

We also face the awkward fact that especially wide conduits occur in woody vines (lianas), with diameters sometimes exceeding 300 micrometers. But unlike trees, vines need not support themselves, and their dry densities are concomitantly low. Xylem, we remind ourselves, is wood, both a conductive and a supportive tissue. Perhaps relaxation of their supportive function at least in part underlies the large size of these conduits. And that suspicion points to mechanical support as the main limitation on height.

Before dismissing hydraulics, though, we should note another way it might bear relevance. Niklas and Spatz (2004) have related both maximum tree height and the basic 2/3-power scaling to the problem of supplying an ever-increasing overall leaf area with water—an argument based on supply rather than pumping cost. I like their rationale but remain a bit skeptical. The quantities of water that trees raise and transpire are almost as impressive as the pressures against which they do so. But these quantities far exceed what gets used in photosynthesis, and they vary widely. Nobel (2005) notes a forty-fold range in water use efficiency—rate of water use divided by rate of carbon fixation. Furthermore, just as with pressure, the most extreme values (here low ones) come from plants living in dry habitats rather than from especially tall trees.

In short, the original question remains. We may even be looking at the wrong variables. In trying to choose between two different routes through which gravity might affect tree height, we presumed a gravitational limit. Even that presumption may be suspect. First, healthy trees rarely fail by gravitationally driven mechanical collapse. (Where I live, occasional windless ice storms do cause gravitational failure, but the large ones for the most part just lose a few branches.) Second, the correspondence between conduit size and flow speed and acceptance of a considerable rate of cavitation suggests that still wider conduits could be tolerated—conduits such as those of lianas. Finally, the fact that negative pressures at tree-top level exceed, sometimes by large factors, the sum of both gravitational and flow-induced pressure drops suggests that still greater losses from these quarters could be tolerated if need be.

So, perhaps the limit on height might sometimes come from something other than gravity. Trees most often blow over in storms by uprooting, less often by snapping their stems near their bases, and still less by shear-induced snapping higher up. Whichever way, failure most likely results from drag, acting on the crown; the taller the tree, the longer the lever arm and the greater the turning moment. In this scenario, the lateral drag of the crown, mainly due to its leaves, imposes the critical disadvantage of height. Several structural features of leaves and trees (and bamboo culms, etc.) make functional sense as devices to reduce vulnerability to drag, often termed "wind throw." The apparent ubiquity and phylogenetic convergence of leaves that take on low-drag configurations in strong winds (Vogel 1989) suggests that drag surpasses gravity as a hazard.

In particular, the commonness of uprooting implies that much of the problem of a tree must come from a peculiarity of its substratum, the limited resistance of soil to tensile forces. Shear and compression soil can resist, but many, perhaps most, trees may not be able to pull on the ground with particular effectiveness. At one time (and maybe somewhere still) large stumps were pulled directly upward by teams of horses solely with the aid of simple windlasses on superstructures that could be moved from stump to stump.

Trees might stay upright in winds in several ways (Vogel 1996a).

1. With a long, stiff taproot that extends the trunk downward (plate 7.1 left) a tree can take advantage of the shear and compression resistance of soil. If lateral roots near ground level fix the location of the base of the tree, blowing the trunk one way asks that the taproot be forced the other way, compressing and shearing soil. The array of smaller, vertical "sinker" roots from larger horizontal ones may work the same way, as well as providing significant tensile resistance to uprooting through shear numbers and area covered. That combination of taproot, laterals, and sinkers seems to be central to the support system of many trees (pines, paradigmatically) of temperate and boreal forests, trees whose trunks obviously bend in winds.

2. Some tension resistance in the most superficial soil layer can come from the tangle of the roots of surrounding vegetation, something of which many tropical trees take advantage with large, thin, upwind buttresses (plate 7.1 right). These act like diagonal cables from trunk to roots rather than the compression-resisting buttresses of Gothic architecture—the misleading linguistic analogy tends to confuse things (Smith 1972; Ennos 1993a). Again, sinker roots assist. Trees with such tensile buttresses are usually thin relative to their heights.

3. Ground-level lateral extensions of the trunks of many big temperate-zone broad-leaved trees are lower and thicker; they most likely work as conventional downwind buttresses that take advantage of soil's reliable compression resistance—as well as providing attachment points for sinker roots. Trees with these wide, heavy bases ("plates" sometimes) typically have thick trunks of dense wood that do not bend noticeably in winds. The arrangement comes into use as a tree matures and shifts from system (1) to system (3).

The wide bases and stiff trunks of system (3) may carry another message. I've just suggested that the vulnerability of trees to windthrow shows that gravity isn't the physical agency that inevitably limits height—whether through fluid or solid mechanics. Compressive buttressing and thick, stiff trunks suggest that gravity may at times operate on the other side of the equation, assisting a tree in staying erect. When pine trees, with little buttressing, blow over, their trunks lie directly on the ground (plate 7.2 top). By contrast, when trees such as large oaks blow over, the bases of the trunks often lie one or two meters above the ground (plate 7.2 bottom). In uprooting, such compressively buttressed trees pivot around a horizontal axis well to the side of the axis of the trunk, as in figure 7.4. To make such a tree uproot, the turning moment must exceed the stabilizing moment. Thus, the product of drag times the height of the center of the crown must exceed the product of the weight of the tree times the distance from trunk axis to turning axis. That simple view ignores any contribution from soil around the roots, of sinker roots, and so forth. But it exposes the possibility that a tree might use its weight to stay upright, with its sinker roots mainly serving to keep it from sliding sideways.

Does such a model survive quantification? Consider a tree with 30 meters of cylindrical trunk, 0.7 meters in diameter, of a density of 1000 kilograms per cubic meter, a pivot point 1.5 meters to one side of the trunk's vertical axis, an otherwise weightless basal plate, and a weightless, spherical crown of branches and leaves. Using the symbols of figure 7.4, the stabilizing moment will be

$$\rho_{tree} \pi r_{tree}^2 h g r_{base}. \tag{7.11}$$

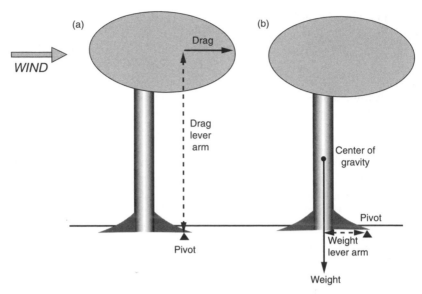

Figure 7.4. The drag of a wind loads a tree not as a column but as an end-loaded cantilever beam. A tree with stiff trunk and basal, compression-resisting buttresses will suffer "wind-throw" when the turning moment from drag and the height of the crown (a) exceeds the opposing moment from its weight and the width of the buttressing (b).

The tipping moment will be the drag of the crown times the height of the tree,

$$0.5C_d\rho_{air}\pi r_{crown}^2 v^2 h. \tag{7.12}$$

Assuming a drag coefficient, C_d, of 0.1, appropriate for a large sphere in fast flow, an air density of 1.2 kilograms per cubic meter, and a speed of 35 meters per second, we equate (7.11) and (7.12) and solve for the radius of the crown. It comes to almost 5 meters, a diameter of nearly 10 meters. While perhaps a little smaller than most real trees, it's close enough to suggest taking seriously this model of an oddly detached tree.

Still, I emphasize its crudeness. We have distressingly little information on the real drag of this kind of broad-leaved tree in high winds. I did some work on the drag of individual leaves and small clusters (Vogel 1989), enough to undermine confidence in any extrapolation or estimate for whole crowns, something Ennos (1999) has reemphasized. Besides the obvious logistical problems, people who run sufficiently large wind tunnels do not take kindly to tests of items expected to fail by detaching pieces that are located just upwind from valuable and vulnerable fans and motors.

Consider what this model says about the relevant variables. First, wind speed has a severe effect on the result (as observed!). Second, height does not directly matter, since it equally affects the weight of the tree and the moment arm of its drag. Greater height does, though, require that the trunk be wider to have the additional flexural stiffness needed to minimize lateral movement of its center of gravity. Of course wider means heavier and thus gives further improvement of a tree's stability. Finally, gravity itself aids stability, as in equation (7.11), so if gravity were greater, such a tree might be able to grow taller—unless, as suggested in the last chapter, air density (and thus drag) were thereby also increased. But whatever the specific value of g, in this model the tree depends on gravity to stay erect.

Any limitation on height must most often operate through the competitive interactions of individual trees. If height does scale with diameter to the 2/3 power and thus cross-section to the 1/3 power, then successive increments in height demand making ever-increasing amounts of wood. Better access to sunlight than one's peers extracts an ever-increasing constructional penalty. In addition, extending significantly above canopy level should disproportionately increase peak wind speeds and thus drag. So any cost-benefit analysis ought to include competitive interactions and growth. And growth depends on a host of other factors. For instance, the dipterocarp forests of Southeast Asia, growing on rich, volcanic soils, achieve greater canopy height than tropical forests elsewhere on earth. Givnish (1995) expands on this kind of argument, noting the ever-decreasing ability of a tree in a forest to compensate for cost with increased leaf area.

I must admit some attachment to a picture that emphasizes the lateral force of wind, a bias stemming from my own interest in air flow and drag. So I hasten to remind the reader (and myself) of the old adage that when one's tool is a hammer, all problems resemble nails. It well may be a case, as said of raccoon- and opossum-hunting dogs in this part of the world, of barking up the wrong tree.

THE DIVERSE ROLES OF GRAVITY

In aerial systems, gravity impels dense bodies downward, with only the relationship between size and descent speed at all negotiable. In terrestrial systems gravity may be less insistently intrusive, but it plays a wider range of roles. Here we moved from cases where the role of gravity was straightforward to ones of increasing subtlety—clearly important, but in ways that challenged our analyses. But I conclude with a mild caution, pointing out that many other cases might have been considered, that this chapter just scratches the surface. It might have compared impact loading with gravitational loading in various forms of locomotion. It might have noted the shift in mammalian posture from flexed-legged to straight-legged, a likely consequence of the way body weight scaled with volume while postural muscle force scaled with cross-section. Or it might have suggested that an alteration of gravity's strength (or wood's strength-density relationship) would affect the length and taper of branches more than it would the overall height of trees.

In this book I've made much of scaling rules and their particular exponents; the way blood pressure depends on body size illustrates one hazard of the approach—such a threshold effect that would be missed by the normal regression-based scaling analysis. For gait transitions we do have a scaling rule, based on Froude number, but here the rule itself applies to thresholds. For tree height, we examined the near constancy of forest heights over space and time, suggestive of mechanical (solid or hydraulic) limitation. Not only couldn't we pinpoint the limitation, but we couldn't either confirm or discredit a scaling rule—or even fully convince ourselves that gravity contributed to the limit.

CHAPTER 8

Gravity and Life in Water

HYDRO-STATIC

Life was born in water, and aqueous habitats still hold most of life's diversity. The nearly aqueous density of most organisms ensures something close to suspension by the surrounding water. A creature might be twice as dense as the medium but never, as on land or in the air, a thousand times as dense. Gravity? We might expect only a minimal impact on design and deportment. But, as we'll see, aquatic life's similarity in density to its medium misleads us.

As mentioned in the last chapter when considering the ascent of sap in trees, in aquatic systems gravity induces a change in hydrostatic pressure with height or depth of roughly 10,000 pascals per meter—an atmosphere for every 10 meters. How potent must that gradient be in oceanic water columns of hundreds or thousands of meters! That hydrostatic squeeze will mercilessly compress a bubble of air or any other gas. Gases, whether pure or mixed, follow Boyle's law of 1662, the law that volume varies inversely with pressure. A bubble of air at a depth of 10,000 meters, that of deep ocean trenches, will have only about 0.1 percent of its volume at the surface—the thousandfold pressure increase will result in a thousandfold volume decrease. Unless in air-saturated water or impermeably encapsulated, the bubble will in short order redissolve, now a victim of two nineteenth-century laws, Henry's and Laplace's. Henry's law declares that increased pressure leads to increased solubility of gases in liquids; Laplace's law (as we now know it) says that the smaller a bubble, the greater the internal pressure from the squeeze of surface tension. Maintaining a gas under water thus bumps into the twin difficulties of depth-dependent volume (Boyle, augmented by Laplace) and dissolution rates (Henry).

That implies major effects of gravity on aquatic life. But we can easily be misled. What about a bubble of some liquid, perhaps a vacuole of lipid? Or a cell, separated from the ocean by a lipid membrane? Or some solid material such as bone or chitin? For liquids and solids, no analogous rule links pressure and volume, and their responses diverge dramatically from those of gases. Pressure increase produces only minor volumetric change. The descriptive variable here (lacking a general rule) is the bulk modulus,

K (or its reciprocal, the compressibility). K is the ratio of change in pressure, Δp, to change in volume, ΔV, relative to original volume, V_o:

$$K = -\frac{\Delta p}{\Delta V/V_o}. \tag{8.1}$$

Minor change means that liquids and solids have very high bulk moduli. Still, while one hears that water is incompressible, the implied infinite modulus does exaggerate. Freshwater has a bulk modulus of about 2.1 gigapascals, seawater about 5 percent more. (Sources vary on the next significant figure.) So seawater is about 4 percent denser at the bottom of a deep ocean trench than at the surface. For liquids, these values are ordinary. The bulk moduli of pure hydrocarbons (octane, for instance) run about half water's value, but such oils as cells might put in vacuoles (vegetable oil, in one tabulation) differ little from water. Solids run one to two orders of magnitude higher, which is to say that they compress even less easily—glass has a bulk modulus of about 40 gigapascals and steel about 160 gigapascals. Even allowing for some pressure-dependent variation of values, in its hydrostatic manifestation gravity should matter little to either liquids or solids.

Pressure exerts slightly more influence on chemistry. At a depth of 10,000 meters, altered hydrogen bonding of water increases its dissociation constant, for instance 2.5-fold at 20° C (Hills 1972). So at extreme depths life faces significant—but not overwhelming—changes in buffering, protein configuration, membrane permeability, and so forth. DNA remains stable at up to ten times the 1000-atmosphere pressure of the extremes of depth; less barostable proteins still denature only slightly at such pressure (Suzuki and Taniguchi 1972). By comparison, the fall in temperature with depth (with resultingly lower metabolic rates) causes substantially greater changes.

What about the surface of a pond or ocean, with air above and water below? Gravitational effects can range from profound to trivial. A liquid's surface prefers to be horizontal and smooth. Gravity provides the main impetus for both, but smoothing involves another agency as well. Surface tension demands that work be done to create additional interfacial area, so it contributes another smoothing force. Disturbances of the smooth surface propagate as waves, and both surface tension and gravity determine their behavior. For water beneath air, surface tension sets the predominant rules for what we call "capillary waves," those with wavelengths below 17 millimeters. Gravity matters most for those above 17 millimeters. Among capillary waves, those of shorter wavelength propagate faster; for gravity waves shorter ones propagate more slowly. Before dismissing capillary waves as relevant only to whirligigs and water striders, bear in mind that every big wave started small, with an initial wavelength at which surface tension ruled.

Nonetheless, gravity does afflict most physical phenomena at the interface between sky and sea. I mention capillary waves both as a caution before putting that interface aside for another occasion and as a way to bring up surface tension—at this point I'll look only at events well beneath the water's surface. Even so, the phenomena that follow must be viewed as an idiosyncratic selection. No space will be given to density gradients caused by depth-dependent changes in temperature or salinity and thus to thermoclines and salt-wedges. Nor to the depth limitations of a chest-powered breathing snorkel, nor to the increasing effectiveness with depth of suckers such as those of an octopus. Attention will be limited to a few interrelated situations—problems of handling undissolved gases and of ballast and buoyancy control.

Of particular interest as we will see are the diverse instances in which submerged organisms maintain stores of air or other gases. They do so for either (on occasion, both) of two main reasons. For some, the air-breathers, what matters is the gaseous oxygen in the mix. For others a gas counteracts body densities greater than that of the surrounding water—gas for flotation. Organisms may store gases internally, in cuttlebone and diverse bladders, or externally, as bubbles or body sheathing. Gas stores may be long-lasting or demand periodic replenishment, the latter from de-dissolution (secretion) or transport downward from the surface. Only in the shallowest and most turbulently moving water will aquatic life be surrounded by water that's gas-saturated at local depth rather than containing gas equilibrated with the atmosphere above the surface. Aquatic organisms containing gas can face Henry's law from either direction. Sometimes gas must be kept from disappearing into solution lest an air breather sink or asphyxiate; sometimes gas must be kept in just such solution lest it tear up tissue or impede circulation.

A final prefatory note—in moving from the aerial and surface worlds of the previous chapters, the practical problems (locomotion aside, of course) shift from forces and accelerations to changes in pressures, volumes, and solubilities.

USING SURFACE TENSION TO EXTRACT GASES FROM WATER

Surface tension can provide the functional equivalent of waterproofing, doing the job well enough to prevent bulk gas loss from a bubble and leaving dissolution as the remaining concern. In my youth, I learned that one could knot each leg of a pair of pants, wet the fabric, and use the pair, held upside down, as a float—not that I ever knew anyone driven to do it. (For an easy test of the trick, try a pillowcase.) Either air or water passes through the pores in the fabric with little resistance, but the interface between them cannot do so, at least if the fabric contains little residual laundry surfactant. At least one spider, *Argyroneta aquatica*, which makes its web within the submerged vegetation of ponds, maintains an analogous air store. Like other spiders an air-breather, it fills a silken bell (that looks like an old-fashioned diving bell) with air that it carries down from the surface. To offset both oxygen use and dissolution, it periodically adds air to the bell.

How fine must the web's mesh be to prevent escape of air, assuming (as seems to be the case) high hydrophobicity? Surface tension (0.073 newtons per meter in freshwater at 20°C) keeps the air contained; hydrostatic pressure forces the air upward through the mesh. So the tighter the mesh (lower d), the deeper the spider can dwell. For a spherical shell whose radius of curvature is r and corresponding diameter d, the pressure developed by surface tension (γ) is

$$\Delta p = \frac{2\gamma}{r} = \frac{4\gamma}{d}. \tag{8.2}$$

Hydrostatic pressure, of course, is simply

$$\Delta p = \rho g h, \tag{8.3}$$

where ρ is the density of water and h is the depth beneath the surface. The depth limit must be the level at which the second reaches the first, so

$$d = \frac{4\gamma}{\rho g h}. \tag{8.4}$$

For depths up to 10 centimeters, the strands must be less than 0.3 millimeters apart—not a particularly daunting requirement. Neither its overall curvature nor the volume of air in the bell make a difference. (Setting up a dimensionless ratio of the pressures by dividing equation 8.2 by equation 8.3 gives an index for the practicality of using surface tension to maintain air under water, the reciprocal of the Bond and Baudoin numbers mentioned in chapter 6.)

Equation (8.4) shows how for a given surface tension, density, gravitational acceleration, depth, and mesh size vary inversely with one another. Of course dissolution rate cares nothing about mesh size and will increase with depth; dissolution rate and the distance air must be transported downward probably limit the arrangement more than does web mesh. Schuetz and Taborsky (2003) note that excessive buoyancy affects these spiders when they bring air down to their bells, so transport looks costly or at least bothersome.

Many insects do much the same thing, if on a less impressive scale. Virtually all adult insects and many of their aquatic larvae, nymphs, and pupae require access to gaseous as opposed to dissolved oxygen. Many hold air bubbles, periodically renewed by trips to the surface, in their various external irregularities—between body segments, beneath wings and elytra, and so forth.

A serendipitous physical phenomenon increases the persistence of such bubbles. Even in water that flows almost unnoticeably (chapter 1), solubility rather than diffusion coefficient determines how fast gases diffuse in or out of a bubble. And much less nitrogen than oxygen dissolves in a given volume of water (1.7 percent versus 3.5 percent by volume at 15° C and 1 atmosphere; Krogh 1941). A bubble thus loses nitrogen more slowly than oxygen. Since air is mostly nitrogen (about 80 percent), the presence of nitrogen thus increases the persistence of a bubble relative to one of pure oxygen. In a classic experiment, Ege, in 1918 (cited by Thorpe and Crisp 1947), found that water bugs with air-filled bubbles could manage seven-hour submergences. By contrast, bugs with bubbles of pure oxygen lasted only thirty-five minutes as outward dissolution, far more than consumption, rapidly decreased bubble size.

Could the respiratory depletion of an external air store be offset by net inward diffusion or by some other device instead of by periodic renewal at the surface? Harpster (1941) and Brown (1987) suggested that photosynthetically produced oxygen might be acquired from aquatic plants, but no quantitative investigation seems to have been done. But at least two ways can demonstrably compensate for outward diffusion and respiration. One depends entirely on surface tension and was established in a series of papers by Thorpe and Crisp (1947) (and the ones immediately following) and Thorpe (1950). The other, less common, is a hydrodynamic consequence of Bernoulli's principle. Stride (1955) showed how it works.

Surface Tension and Plastrons

First, consider an odd role of surface tension. We've long known that a thin film of air covers some submerged adult insects, a layer conspicuous as a silvery sheen of the sort one sees on a suddenly submerged leaf of nasturtium or lotus. Evidence for its respiratory role goes back at least to Comstock (1887), and Harpster (1944) proved that some insects could maintain the film without periodic replacement, at least in water with dissolved gases at atmospheric partial pressures.

The mechanism depends on the relationship expressed by equation (8.2). Usually we apply the equation to bubbles convex on the outside—as they normally are. The smaller the bubble, the greater the component

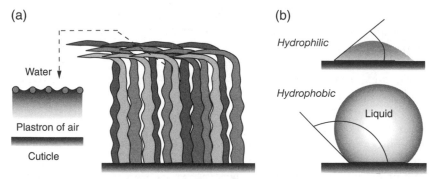

Figure 8.1. (a) Diagrammatic representation of a plastron; (b) contact angles for an aqueous liquid on hydrophilic and hydrophobic surfaces.

of pressure inside caused by surface tension. So tiny ones can spontaneously disappear as that pressure drives their gaseous contents into solution and their size down into Laplace's embrace. For this thin film of air, or "plastron" (Thorpe and Crisp 1947), one has to look at what equation (8.2) says for an anti-bubble, one concave rather than convex on the outside. Smaller (as radius of curvature) now implies lower, not higher, internal pressure. With sufficiently small and numerous bubbles of this sort, enough oxygen to sustain an air-breathing animal might diffuse inward.

So a submerged adult insect needs a lot of tiny, concavely curved bubbles or else some kind of air layer with concave interfaces. These insects create such layers by forming air-water interfaces at the periphery of a dense layer of short, hydrophobic, cuticular hairs, as in figure 8.1a. Thorpe and Crisp (1947) estimated that *Aphelocheirus*, a naucorid bug, had about 2,000,000 hairs per square millimeter—thus individual hairs were less than 1 micrometer apart.

Using newer and better imaging equipment and the same species, Hinton (1976) revised that to 4,000,000 hairs per square millimeter, with hairs tapering from 0.4 to 0.2 micrometers in diameter and extending 3 micrometers outward from the body. As in the figure, their distal portions are bent parallel to the surface, with little space between them. Treating the erect parts of the hairs as columns with the typical stiffness of arthropod cuticle (10 gigapascals) and assuming that a hair is vulnerable to Euler buckling, Hinton (1976) calculated that buckling such an array of hairs would require a pressure of about 40 atmospheres. So the buckling strength of the hairs would impose a depth limit for use of a plastron of 400 meters, no problem for an entirely freshwater fauna. In practice,

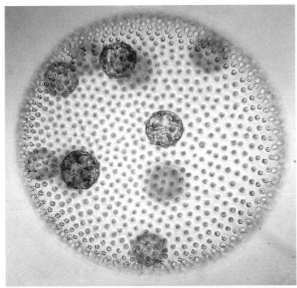

Plate 1.1. *Volvox*, a colonial alga, with daughter colonies within. *Chlamydomonas*, which will appear later, is a closely related unialgal form (courtesy of Aurora Mihaela Nedelcu).

Plate 1.2. Two inhabitants of freshwater ponds and lakes—*Spirogyra* (*left*) and *Ranunculus fluitans* (*right*), the latter quite similar to *R. pseudofluitans*, courtesy of John Somerville.

Plate 2.1. An exploded fruit (with seeds) of *Hura crepitans* (© 2008 Wayne P. Armstrong); the inset shows an intact fruit (courtesy of Organization for Tropical Studies/La Selva Digital Flora Project).

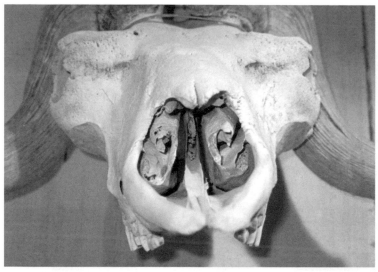

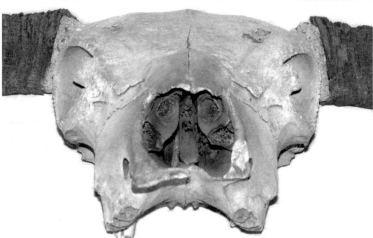

Plate 9.1. Front views of the skulls of a musk ox (*top*) (at the Musk Ox Farm, Palmer Alaska) and a domestic cow (*bottom*) (at the Museum of Life and Science, Durham, North Carolina).

Plate 9.2. *Left*, a turret spider, probably *Geolycosa pikei*, from Bogue Bank, North Carolina; *upper right*, a burrow, with as minimal a turret as one finds; *lower right*, a spider that, when cool, has been exposed to warm, moist, moving air—unevenly dusted with uranine (sodium fluorescein) and illuminated with ultraviolet light.

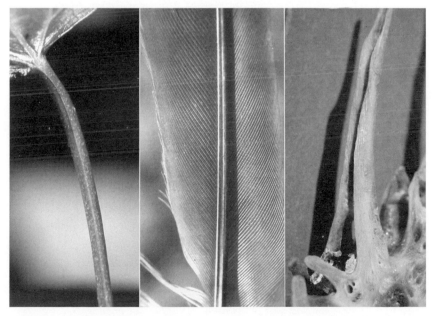

Plate 11.1. Some biological beams with longitudinal grooves. *Left*, petiole of sweetgum (*Liquidambar styraciflua*); *middle*, feather shaft of blue jay (*Cyanocitta cristata*); *right*, neural spine of an unidentified bony fish.

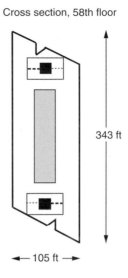

Cross section, 58th floor

343 ft

← 105 ft →

Plate 11.2. *Left*, the John Hancock Tower, in Boston, as seen from MIT, in Cambridge. *Right*, a diagram of the damping system—two rooms house the 300-ton sliding blocks, each attached to the adjacent walls by a spring and a damper.

Plate 12.1. *Left*, the lower part of a willow oak (*Quercus phellos*) grown in a fairly exposed location. *Right*, the lower part of a baldcypress (*Taxodium distichum*) in the shallow water of a coastal swamp in North Carolina. (© 2008 Mary Henderson.)

imperfect hydrophobicity and consequent failure of surface tension failure restricts plastron use to lower hydrostatic pressures and shallower
depths. Both Thorpe and Crisp (1947) and Hinton (1976) found that
plastrons break down through wetting at about 3 atmospheres above
ambient pressure. Still, 3 atmospheres corresponds to a depth of about
30 meters, fairly deep by the freshwater standards of insects.

The relative role of the geometry of the outer part of the plastron and
of its wettability remain uncertain. The usual measure of hydrophobicity
is the angle between the surface of a bubble and the surface it contacts, as
in figure 8.1b; an angle of 180° would indicate perfect hydrophobicity.
The hairs (and cuticle generally) cannot be perfectly hydrophobic—we
know nothing that extreme. Contact angles for waxy coatings range between 105° to 110°, the range assumed in most calculations of plastron
performance. More recent work, though, has uncovered biological systems with higher angles. Water on the surfaces of the leaves of lotus and
some other plants that have complexly sculptured waxy cuticles can
reach 160° (Barthlott and Neinhuis 1997; Neinhuis and Barthlott 1997).
Water droplets roll around on them as dirt-scavenging beads (plate 8.1).
Of especial relevance here, Wagner et al. (1996) reported angles as high
as 155° on insect wings, and Gao and Jiang (2004) reported an angle of
168° for the legs of a water strider—an insect for which high hydrophobicity should be particularly advantageous. These high values depend on
surfaces with roughness of the same scale (Feng et al. 2002) as the conspicuous humps on the plastron hairs in Hinton's (1976) scanning elec-
tron micrographs.

Plastrons turn out to be widespread among arthropods and have undoubtedly evolved many times, perhaps because a hydrophobic exoskel-
eton with minute outgrowths represents nothing out of the ordinary and
because they include nature's most diverse air-breathers. Plastrons occur
among eggs and larvae that suffer occasional floods (Hinton 1976), and
they appear as well in some millipedes, mites, and whip scorpions (Hebets and Chapman 2000). And interest in them has become increasingly
widespread. Bush et al. (2008) provide a general view of the mechanisms
of hydrophobicity (and superhydrophobicity) and its use in arthropods.
And Flynn and Bush (2008) give a careful and thorough review of both
plastrons per se and the various other ways of maintaining air
underwater—with an eye to defining the options for practical bits of human technology.

The self-cleaning, buoyancy-augmenting, and protective roles of super-
drophobicity give it considerable technological appeal; Solga et al. (2007)
explore both the present state-of-the-art and the various as-yet-unrealized
possibilities. One product at least has reached the market, a self-cleaning

biomimetic super-hydrophobic coating, "lotusan" (Sto Corp., Atlanta, Georgia, USA).

In the original version of this chapter, I wondered about plant leaves, most of which have hydrophobic outsides and many of which have fuzz as well as stomata on their undersides (or on both surfaces), but I know of no data indicating any analogous functional arrangement, either routinely or during occasional immersions. Plastrons have now been shown both functional and persistent on the surfaces of several species of wetland plants. Colmer and Pedersen (2008) tested submerged leaves, treating a parallel set with a detergent to remove the shiny film of gas. With a gas film, leaves exchanged gas through stomata, as do terrestrial leaves, rather than through the leaf cuticle in the manner of film-less submerged leaves. The film greatly increased both CO_2 uptake in the light and O_2 uptake in the dark. (I'd pat myself on the back were it not for the biomechanical hazard of the maneuver.)

I also wondered whether diffusion alone can adequately transport oxygen to the spiracles through a gas layer 3 micrometers thick or whether some additional physical device awaits recognition. (Chapter 1 raised the possibility of an analogous transport limitation within leaves.) Diffusion-augmenting bulk gas motion within the plastron might be induced by movement of an insect through the water or by such things as local water pumping by hindleg motion, the latter reported by Harpster (1941) long ago. Since then the issue has been addressed in quantitative terms by Flynn and Bush (2008), again defining the realm of the possible.

USING FLOW TO EXTRACT GASES FROM WATER

Far less common than a plastron as a way to maintain air under water is flow-induced local pressure reduction. Only one case has been well documented, a few others remain conjectural; as we will see, all too few situations meet its physical requirements.

In addition to hydrostatic pressure, the surface of an object in a flowing fluid feels the pressures of that flow. The specific pressure at a place on the surface depends on its location, with the maximum at some upstream point. There the pressure equals the local hydrostatic pressure plus a component from local conversion to pressure of the flow's kinetic energy. Bernoulli's principle gives the pressure increase over the local hydrostatic pressure at that point as

$$\Delta p = \frac{\rho v^2}{2}, \tag{8.5}$$

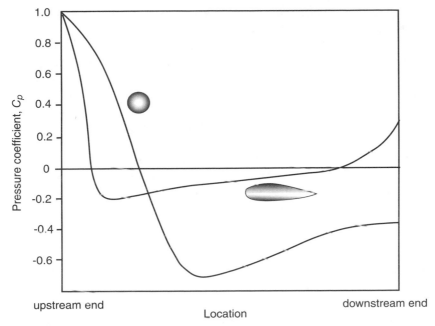

Figure 8.2. Pressure coefficient versus location on the surfaces of two objects—a sphere and a streamlined body of revolution. Both are about 40 millimeters in diameter. The data come from tests in air at 10 meters per second and thus a Reynolds number (based on diameter) of 25,000.

where v is the speed of flow before slowing by the object. Downstream, pressures are inevitably lower, with specific values determined by location and the object's shape.

These downstream pressures (Δp's) are commonly expressed as their ratios to that maximum; the resulting dimensionless variable goes by the name "pressure coefficient," C_p, so

$$C_p = \frac{2\Delta p}{\rho v^2}.$$
(8.6)

Thus a graph of pressure coefficient versus distance on the surface from upstream to downstream must always start at the y-axis with a value of $C_p = 1.0$, as in figure 8.2. In effect, pressure coefficients adjust pressures for the effects of speed and fluid density, dedimensionalizing them.

Not only does the pressure coefficient never reach 1.0 anywhere downstream, it drops below ambient over much (usually most) of the rest of the body. The positive region turns out to be surprisingly limited, not even extending back to where the body is thickest. (For unstreamlined objects the pressure coefficient remains below zero back to the rear end, while for streamlined objects it gradually returns to positive territory, eventually approaching but never quite reaching 1.0.) Of present relevance, the overall pressure coefficient, integrated over the entire body, will almost always be negative—that is, outward; the particular value depends mainly on the body's shape. As a result, a bubble held stationary in a flow develops a net outward pressure coefficient. That coefficient ought to lie between about −0.1 and −0.3, with more negative values for broader and less tapered bubbles. Again—since it seems paradoxical—the pressure inside will drop below the pressure outside (Vogel 1994b).

(By integrating over the surface, taking local surface orientation into account, one can calculate drag, the downstream force on an object. Here we're concerned, instead, with transmural pressures.)

Flow-induced pressure reduction thus provides another way to extract gas from solution, one whose operating range can be defined quite simply. Assume, as is common in rapidly moving, shallow water, saturation with air at atmospheric—or surface—pressure. If the flow-induced pressure decrease in a bubble exceeds the local hydrostatic pressure increase due to depth, then the bubble should act as a gas extractor. So the possibility of gas extraction can be specified in terms of the ratio of equation (8.3) and equation (8.6),

$$\frac{\Delta p_{flow}}{\Delta p_{hydrostatic}} = \frac{-C_p}{2} \frac{v^2}{gh} > 1.0. \tag{8.7}$$

(The minus sign on the right acknowledges the comparison of a pressure drop with a pressure increase.) Incidentally, the dimensional variables in the ratio form the Froude number, the ratio of inertial force to gravitational force, mentioned in the previous two chapters. Here it appears in the guise of flow force over hydrostatic force.

But equation (8.7) defines a daunting condition. A bubble must be maintained despite the drag of the flow around it. And flow works far less effectively than surface tension as a pressure-reducer, at least comparing typical flows with surface tension acting across severely curved interfaces. So the bubble must be held by an animal in the face of a substantial current, and it cannot be far from the surface. Figure 8.3 defines the limits for three possible pressure coefficients. It suggests that at a current speed of 1 meter per second a bubble would have depth limits between about 5 and 15 millimeters. Worse, given the inevitable velocity gradients near surfaces, an organism holding onto some solid surface may be exposed to a local flow substantially slower than that of the mainstream. Were it not for one well-documented case, we might dismiss the scheme as creative but impractical. That case merits some attention.

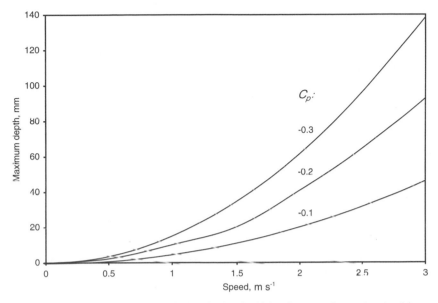

Figure 8.3. The maximum depth at which a bubble of air can be maintained by flow-induced gas dissolution for three different pressure coefficients. The water is assumed saturated with air at atmospheric pressure.

Working in Ghana, Stride (1955) noticed that adults of a particular kind of elmid beetle, *Potamodytes tuberosus*, often "appeared to fly straight into the rushing water" of a rapid stream and then congregated on rocks just beneath the surface. Each faced upstream and carried "a large silvery air bubble." In the laboratory, its bubble persisted indefinitely if—and only if—rapid, shallow flow enveloped the beetle. With some difficulty, Stride managed to measure the pressure within bubbles on restrained beetles subjected to a range of flow speeds. At the test depth of about 10 millimeters, bubbles persisted at speeds above about 0.8 meters per second.

I've reanalyzed his data, extracting the flow-induced pressure reduction from the background hydrostatic pressure of 98 pascals (that 10-millimeter depth). Figure 8.4 gives his thirty-four points in contemporary units. They don't correspond to a specific value of pressure coefficient: the faster the flow, the lower its apparent value. That comes in part from change in bubble shape with flow speed, as he notes, but changes in the air-water interface just above the beetles probably contribute as well. A linear regression nicely fits his data and as nicely misleads, by extrapolation implying an impossible flow-induced pressure drop of 69.3 pascals with no flow at all. Nonetheless, as the figure shows, the data do correspond to reasonable

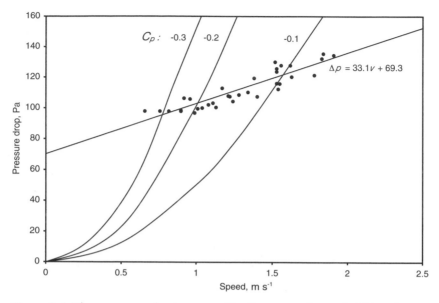

Figure 8.4. The pressure reduction caused by flow around a gas bubble—the predictions for three pressure coefficients and the measurements (points and fitted equation) by Stride (1955).

values of pressure coefficient, lending credence both to his measurements and to the present analysis.

Yes, the conditions may be daunting, but Stride noted that such beetles were common enough in his area, and Brown (1981) believes that elmid beetles of the genera *Hispaniolara* and *Potamophilops* play the same game. And elmids are not the only kind of riffle beetle. Moreover, adult riffle beetles use plastron respiration, so they already have spiracular connections to an outer store of air (Brown 1987). Stride's beetles show no other obvious anatomical adaptations, which implies both that such a trick can be done with behavior alone and that lurking cases might easily be missed. Finally, subsurface photosynthesis (or dissolution of entrained bubbles from local waterfalls) can raise the partial pressure of oxygen beyond atmospheric level (Giller and Malmqvist 1998), which would allow a bubble to be maintained to greater depth—although Stride excluded the possibility in his particular system. Perhaps mere inattention explains why no others have surfaced in the intervening half-century. The composition of a diverse but globally consistent "torrential fauna" has been well-studied, although more with regard to who lives where than to any functional issues. The older accounts, especially Hora (1930) and Nielsen (1950), are still worth reading in any search for candidates.

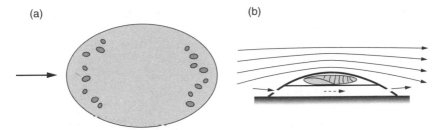

Figure 8.5. The cocoon (about 15 millimeters long) and pupa of a pyralid lepidopteran. (a) Top view of the cocoon, showing upstream and downstream openings. (b) A diagrammatic cross-section of the cocoon, with the pupa resting on a shelf within an air-filled upper chamber (based on a variety of drawings and descriptions, incorporating agreed-upon features).

Beyond the "plunge-and-grab-on" trick of *Potamodytes,* other ways of using current to maintain an underwater store of air must be possible. According to Pommen and Craig (1995), the plastrons of pupal net-winged midges (blepharicerid flies) produce bubbles that then persist in the low-pressure vortices behind their gills. Let me suggest yet another arrangement on the chance that it will either stimulate a specific investigation or consideration of still others.

The larval stages of many pyralid Lepidoptera (especially the Nymphulinae) are entirely aquatic, although pupae and adults typically live in air. In at least some genera—*Aulocodes* in northern India (Hora 1930) and *Elophila* and *Petrophila* (= *Parargyractis,* sometimes *Paragyractis*) in North and South America—the last larval instar spins a tightly woven cocoon atop a submerged rock in rapids. The pupa then rests on a shelf within the cocoon, as in figure 8.5a. The space above the shelf is consistently and persistently air-filled, air of unknown origin as put by Nielsen (1950) and Resh and Jamieson (1988). All descriptions mention upstream and downstream holes that allow ingress and egress of water as well as flow beneath the shelf.

Since the openings are small and close to the substratum, I doubt if water flow through the cocoon would ever be rapid enough for oxygen extraction. But flow across the top could do the job. Pressure will be locally reduced by the velocity increase (shown in figure 8.5b as compression of streamlines) needed to bring water over the cocoon. Animals certainly use the locally reduced pressure as fluid flows over a protrusion for a variety of purposes—it has been shown in systems such as sponges (increasing flow for suspension feeding), keyhole limpets (increasing respiratory water movement), and giant termite mounds (increasing respiratory air movement) (Vogel 1994b). Surface tension between air and water within the outer wall of the cocoon should add the equivalent of a plastron, one in which the low pressure difference would demand only a coarse mesh and weak structure. Unlike the situation in an uncontained bubble, in a plastron a solid structure helps offset inward fluid pressure. That gives relevance to Lloyd's (1914) comment that the outer walls of the cocoon "are spun of thick inflexible layers of silk." Perhaps these lepidopteran pupae combine two mechanisms, the surface tension pressure barriers of plastrons with flow-induced persistent gas bubbles.

That raises the question of whether ordinary plastrons on insects that live in rapid flows take advantage of currents to reduce pressures still further than could surface tension alone—Pommen and Craig (1995) mention the possibility. I'd guess that the effect, while inevitable, must be minor except perhaps in very shallow water. After all, currents that an insect might encounter and withstand will generate at most a few hundred pascals of pressure, while unassisted plastrons can withstand pressures a thousand times greater.

Dealing with Buoyancy

Gravity acts not so much on an organism's mass per se, as on the difference between that mass and the mass of the fluid that the organism displaces. Terrestrial organisms may have densities much higher than that of the air around them, but aquatic organisms differ so little in density from water that their displaced fluid cannot be ignored. Still, that difference varies from organism to organism and even, through both passive and active alteration, for a given organism over time.

Marine organisms without solid supporting materials have densities especially close to that of their surrounding fluid (around 1026 kilograms per cubic meter), so they live in a weightless (although emphatically not massless) world. Minor adjustments in such variables as ionic composition can handle any residual weight (or buoyancy), and they often maintain a slight downward bias. An equivalent freshwater organism can lower its density to that of the surroundings with a small body of some lipid, since lipid densities run around 900 kilograms per cubic meter. (Cholesterol, with a density of 1067 kilograms per cubic meter, cannot be used, while squalene, at 860 kilograms per cubic meter is particularly effective.) Since lipids have about the same compressibility as water, buoyancy does not depend on depth.

Either too great or too little density can make trouble. Thus insufficient density can hamper the ability of sedentary organisms to stay put as water flowing over them imposes both drag and lift. Or it can limit modes of locomotion that depend on surface purchase. The typical remedy for insufficient density consists of adding biosynthesized or environmental stony material to the system. For instance, various insect larvae (trichopterans, most notably) incorporate tiny pebbles into their cases. They reportedly use larger pebbles in swifter flows, although that may just be a result of differences in what's at hand (Pennak 1978).

One can point to many cases of deliberate density increase. A wide variety of air-breathing aquatic vertebrates, both fossil and living, swallow and retain stones ("gastroliths") to offset buoyancy from air-filled lungs—plesiosaurs, some crocodilians, some pinniped mammals, some penguins, and others (Taylor 1993). Marine gastropod and bivalve mollusks commonly have thick shells of calcium carbonate that at least on some occasions must help them hold position in currents. Sessile adult bivalves that lack specific attachment devices (such as the byssus threads that tie mussels to rocks) tend to have the thickest shells. A tiny surf clam (*Donax variabilis*) depends on staying near the substratum as wave swash or backwash moves it up or down a beach. According to Ellers (1995), its relatively thick and dense shell gives it an overall density of 1650 kilograms per cubic meter; the densities of other bivalves, mostly larger, from the same beaches, range from 1170 to 1660.

Density costs little in marine habitats because seawater is usually saturated or supersaturated with calcium, and because stones of greater density than calcium salts are common enough. The more aquatic mustelid mammals seem to have denser bones than the less aquatic ones—but with ambiguous functional significance (Fish and Stein 1991). Good organisms for exploring functional increases in density might be freshwater mussels—all shelled, diverse in size and flow speed preferences, and bottom dwelling but not attached. Many, perhaps most, should be at some risk of dislodgement through sediment erosion during floods. One encounters anecdotal statements about thicker shells beneath more rapid waters (as in Pennak 1978), but no systematic study that rules out some simple scaling rule for size versus density or shell thickness seems to have been done (Barnhart, personal communication).

Alternatively, solid supportive systems can raise densities enough to make pelagic organisms sink. Compensatory adaptations for increasing buoyancy seem to be more common than those for reducing it, perhaps because the big animals we mainly study generally have such hard supportive elements. (Hydroskeletons are the most widespread exception.) Bones and stony corals have densities around 2000 kilograms per cubic meter, mollusk shells of calcium carbonate (as calcite or aragonite) around 2800, and crustacean exoskeletons of calcified chitin about 1900 (Wainwright et al. 1976). In general larger animals devote a greater fraction of body mass to support—even, if less dramatically, among aquatic ones—so the problem gets worse with increasing size.

Some animals take a brute force approach, producing sufficient lift to offset their negative buoyancy as long as they keep swimming. The paradigmatic examples are pelagic sharks, which lack swim bladders. Lift comes from the combined action of asymmetrical caudal fins—larger lobes above than below—and a body pitched nose-upward. The fin asymmetry causes a downward tilt to the tail's rearward force (29° below horizontal in a leopard shark), generating lift as well as the thrust needed for swimming. Trouble from the posterior line of action of that lift is offset by additional lift from a flattened head and upward body pitch (11° in a leopard shark), a combination with an anterior line of action (Wilga and Lauder 2002). Still, sharks minimize the need for lift with skeletons that are much less calcified than those of bony fish of similar size—"bony" recognizes just that difference. And squalene, with that especially low density of 860 kilograms per cubic meter, makes up a large fraction of their lipid; indeed it was named for its common source, shark (L. "squalus") liver.

Squid likewise lack flotation devices to compensate for negative buoyancy, and they also make do with a minimum of stiff material. Their main skeletal element is a light, thin lengthwise "pen" that keeps the upstream (posterior) end from bending when the jet gives a forceful squirt. Since it can direct its jet downward, a squid need not make headway to maintain enough lift. But it must still work—hovering costs about twice as much as does resting and almost as much as does normal locomotion (Webber et al. 2000).

A few bony fishes, members of an obscure group that lacks swimbladders, take the shark game a step further. These Antarctic notothenioids have bones with only a trace of ossification—their ashed skeletons weigh only around 0.4 percent of body weight rather than a typical 2 percent. And they are full of lipid, mainly triglycerides of about 930 kilograms per cubic meter, located subdermally and in intermuscular sacs (DeVries and Eastman 1978, Eastman and DeVries 1982). With near-perfect neutral buoyancy, they need not depend on swimming for hydrodynamic lift.

Using Gases at Local Pressure for Buoyancy

The densest gas weighs much less than the least dense liquid, so gases provide the most space-efficient device for buoyancy augmentation. At atmospheric pressure, air gives 700 times more buoyancy per unit volume than the best lipid, squalene. Even at a depth of 1000 meters, air (or, very nearly, oxygen or nitrogen) will still be about seven times better. And storing air should entail less metabolic expenditure than storing lipids, whose synthesis is especially costly.

Several problems, though, offset that high volumetric efficiency and low cost. As already mentioned, the solubility of gases in liquids varies with pressure, so a quick reduction in pressure may bring dissolved gases out of solution. Human divers ascending too quickly risk "the bends," the name alluding to the stooped posture of those who have repeatedly endured painful gas bubbles trapped in their joints. We can equilibrate with local pressure at depths up to 100 meters or so in a diving bell, caisson, or when using an aqualung. But we then need a slow ascent to allow time for dissolved gas to work its way out through the lungs rather than vaporize within our blood and other tissues. Of the gases in air, nitrogen makes the most trouble, partly because air contains so much of it—80 percent by volume—and partly because of its substantial solubility in blood and tissues and its high solubility in body fat. Diving with helium works better because its lower solubility more than compensates for the greater rapidity with which its smaller molecules diffuse into the body. Schmidt-Nielsen (1997) gives a particularly good account of the relevant physiology.

Nitrogen dissolution causes less trouble for diving animals than for us, mainly because they don't breathe from tanks of compressed air while under water. So only gas already present in their respiratory passages can go into solution. Diving animals usually minimize that volume by exhaling before leaving the surface, tolerating the resulting extra thoracic compression at depth. But not all do so. Penguins inhale before diving; buoyancy demands that they work hard during the initial phase of descent. In compensation, the buoyancy from air in their plumage and respiratory systems speeds upward gliding during the latter part of their ascents (Sato et al. 2002).

Thoracic compression and the peculiar dynamics of descent and ascent in penguins are only part of the problem. Gases compress all too readily, with volumes running almost exactly inverse with local pressure, in sharp contrast with the minor volume changes in other body constituents. Thus at only one depth can an organism containing air at local pressure be neutrally buoyant. Worse yet, that neutrality is metastable. Ascend, and the gas volume and thus buoyancy increase, descend, and buoyancy decreases. Putting the gas in a rigid container, as we do in submarines, would solve the problem, but the necessary stiffness for the container's wall limits that fix. Many kinds of surface-living or rooted aquatic organisms use inextensible gas-filled bladders, but they just need a little buoyancy and don't change their depth much. And inextensibility requires only tension-resisting materials, relatively cheap to make and light in weight. Withstanding compression well below the surface (as elsewhere) is less easy— columns and beams cost more than ropes.

Consider the poor diving duck. It carries air in its plumage, so it floats high in the water, as in plate 8.2 left. Diving just beneath the water's surface means struggling against excessive buoyancy. Deeper dives cost less to sustain—but reaching greater depth takes more time and energy, both precious resources for an actively swimming air-breather (Lovvorn and Jones 1991). A few birds such as anhingas have hydrophilic plumage and need deal only with internal gases. At the surface, the less buoyant anhinga swims with only neck and head exposed (plate 8.2 right) and with, one presumes, greater locomotory cost. In addition, it loses most of the insulating value of the plumage, restricting it to warm waters, and immediately after emersion it cannot readily fly (Hennemann 1982).

Most bony fishes maintain near-neutral buoyancy with a gas-filled swimbladder, either gulping air at the surface or (more commonly) secreting gas from circulating blood. Freshwater fishes have swimbladders that make up 5.5 to 8.3 percent of body volume, while the bladders of those living in the sea occupy 3.1 to 5.6 percent (Alexander 1966). For comparison, a pair of our lungs (our evolutionary homolog of a swimbladder, incidentally) averages about 4 percent of our body volume. So even with their fine volumetric efficiency swimbladders are not tiny organs.

One can demonstrate the problem of maintaining buoyancy by putting a small goldfish in a large glass jar such as a 40-liter carboy. Aspirate the air above the water even slightly and the fish rises abruptly and only slowly readjusts to the lower pressure. It may even belch a bubble (which will help readjustment)—goldfish have a connection, the so-called pneumatic duct, between swimbladder and esophagus. A fish that has readjusted to a lower pressure will sink when atmospheric pressure is restored, readjusting again after a short time. Such readjustment can offset about a meter of depth (10,000 pascals) per hour (Fänge 1983). If you try the demonstration, don't use just any fish—only some, such as salmon, carp (including goldfish), pickerel, and eels, have pneumatic ducts. Others you'll put at risk of a ruptured swim bladder, which cannot be pleasant.

To put a few numbers on the problem, consider a neutrally buoyant fish that lives near the surface and whose swimbladder occupies 5 percent of its volume. If it descends to 10 meters, pressure will double and the volume of the swimbladder will halve. As a result it will have a density about 2.5 percent greater than the surrounding water. That's enough for a bilaterally compressed body to sink further at an appreciable rate even when maintaining its long axis horizontal. If it descends to 90 meters, pressure will go up ten-fold and swimbladder volume down by the same fraction, to about 0.5 percent of the body. Now 4.5 percent denser than the water, it will descend still faster. One need not consult a graph to recognize that the problem of depth metastability will be most severe near the surface. Abyssal fish should be able to ignore most depth-dependent volume change.

But how can a fish maintain a gas mixture in a living bag at severely elevated pressures? The problem, the inverse of the outgassing of the bends, comes from the same high solubility of gases at high pressures. Blood with hemoglobin can transport a lot of oxygen for its volume, but it does so at a partial pressure no greater than that of the oxygen in the water that passes across the gills. Since the oxygen in deep waters has come either from the air above or from photosynthetic activity near the surface, it will be far below local saturation (partial) pressure. That strongly impels bladder oxygen to dissolve, diffuse into tissues and blood, thence move to gills, and thence pass out to sea.

The solution (in both senses of the word) of fishes has two main components. First, fishes restrict the vascularization of the swimbladder to a tiny gas-secreting gland, lining the rest of the bladder with layers of crystalline guanine that make it almost completely impermeable to diffusing gas (Lapennas and Schmidt-Nielsen 1977). Second, they supply the gas gland with blood that has passed through a particularly efficient countercurrent exchanger—a device described in connection with heat conservation in chapter 5. Such an exchanger enables blood leaving the gland with dissolved gases at high partial pressures to lose gas, not to the gills and exterior, but to blood about to enter the gland. In effect, a bag containing oxygen (among other components) at very high partial pressure can be in diffusive contact with blood at nearly the same partial pressure rather than at the lower partial pressures of the ambient water, the gills, or elsewhere in a fish.

Nonetheless, some work does need to be done—a countercurrent exchanger can only minimize losses and secretory costs. The particular trick of bony fishes consists of acidifying the blood as it passes through the gas gland with CO_2 and lactic acid. That reduces the hemoglobin's affinity for oxygen, driving oxygen into physical solution and increasing its partial pressure in the venous blood returning from the gas gland into the exchanger. Even though the venous blood has less oxygen per unit volume than the arterial blood—some has passed into the swimbladder—its higher partial pressure means that net diffusion will move oxygen toward the arterial blood. So oxygen will head back toward the swim bladder even as the blood that formerly held it goes gillward. (Bear in mind that solutions, unlike gases, have partial pressures, sometimes called "tensions," that depend on solubility as well as fractional composition.) This version of an exchanger has been called a "countercurrent multiplier." Schmidt-Nielsen (1997) again gives a succinct description, while Fänge (1983) supplies quantification and the details of the physiological chemistry. Again, the process still requires continuous work—clever machinery can only minimize the task.

Using Gases at Low Pressure for Buoyancy

In the ways they maintain buoyancy, as in so many respects, the cephalopod mollusks show us evolutionarily achievable alternatives to those of our own phylum. As mentioned, squid, like sharks, make do with a minimum of stiff material and with continuous locomotory effort. More remarkable are their kin, the cuttlefish. They demonstrate that gas can be kept at pressures both constant and well below ambient—circumventing both the metastability problem and buoyancy loss due to gas compression. Better, they do these things without sacrificing the ability to adjust gas volume.

While most bony fish put gas in a single chamber, cuttlefish put it in rigid foam—"cuttlebone." With its small and rigid chambers, each about 0.1 millimeter wide and 0.6 high, cuttlebone represents much more material than does a swimbladder. But extracting its main material, calcium carbonate, from saturated seawater should cost little. The compartmentalization allows the cuttlefish to make a nearly incompressible float of lower density, about 620 kilograms per cubic meter, than that of any lipid, if somewhat denser than the gas plus swimbladder wall of a bony fish. Pet stores sell pieces of dry cuttlebone for caged birds, who sharpen their beaks on it, so one can easily acquire a sample of this light but rigid buoyancy tank. Its unusual mechanical properties have attracted attention from people interested in materials (Birchall and Thomas 1983; Gower and Vincent 1996).

In a series of now-classic papers, Denton and Gilpin-Brown (1961 et seq.) and Denton et al. (1961) figured out how the system operates. Neither gas gland nor countercurrent multiplier plays a role. Cuttlefish balance the hydrostatic pressure difference between the surrounding water and the interior of the cuttlebone with a liquid (within part of it) that has an osmotic pressure below that of the blood. Since the liquid has a lower salt concentration than blood, water is osmotically drawn out of the cuttlebone with the same pressure as it's hydrostatically forced into it. As with a swimbladder, maintaining the system takes work—here the osmotic work of extracting Na+ and Cl– to keep the fluid hypoosmotic. The organ as a whole must (and, of course, does) withstand the local hydrostatic pressure. The measured collapsing pressure, 24 atmospheres, comfortably exceeds the hydrostatic pressure, about 15 atmospheres, at the depths at which the animals live.

One comparison of different flotation media comes from calculations of the fraction of body volume that each would require for a standard body density. If cuttlebone, with its density of 620 kilograms per cubic meter, makes up 9.3 percent of the volume of a cuttlefish swimming in seawater of 1026 kilograms per cubic meter (Denton and Gilpin-Brown 1961), then the rest of the body has a density of 1066. Assuming that density for its body (excluding the gas in the swimbladder), a bony fish would need a swimbladder (containing essentially massless gas) of an internal volume of 3.9 percent, within the reported range. What if a creature used lipid for flotation—say fat or oil of 910 kilograms per cubic meter? It would have to devote 27 percent of its volume to this antiballast. Even the best nongaseous flotation material, squalene, would still occupy 20 percent of overall volume. While plausible (some fishes do contain large amounts of oil and small cetaceans have thick layers of fatty blubber), these fractions imply an energy investment comparable to that of all other body components combined. Moreover, the investment cannot be cashed in during starvation without additional locomotory effort to prevent sinking.

CARTESIAN DIVERS

The depth metastability that bedevils a bony fish and that a cuttlefish evades underlies a wonderfully clever device that once tested the dexterity of physiologists. This manometric apparatus could measure such things as the rate of oxygen consumption of invisibly small organisms or their parts. Physicists have long recognized a "Cartesian diver"—even if Descartes deserves no credit. A "diver," a floating body of minimally positive buoyancy, can be made to sink by applying a small pressure to a container of water. One can be assembled from the simplest of everyday items, as in figure 8.6a, and some version graces science classes at diverse educational levels.

Because of its fishlike metastability, one cannot easily make a diver that, unattended, neither sinks to the bottom nor floats to the top. While not the usual point of the demonstration, its incarnation as a measuring device depends on that metastability. As originally described by Linderstrøm-Lang (1937) and shown in figure 8.6b, a glass "diver" exposed to the local pressure contains a small volume of air, a respiring bit of life in water (plus, most often, a CO_2 absorber), and a droplet of oil as a pressure-transmitting seal. A larger, closed container of liquid (usually ammonium sulfate or lithium chloride to reduce gas exchange) envelops the diver.

Initially the operator holds the diver at some arbitrary depth in the container by manipulating the pressure within the container. If the specimen withdraws gas from the air in the diver, the diver becomes denser and plunges. Reducing the overall pressure will persuade the diver to return to the initial depth. Knowing that pressure and the initial volume of the gas

(a)

(b)

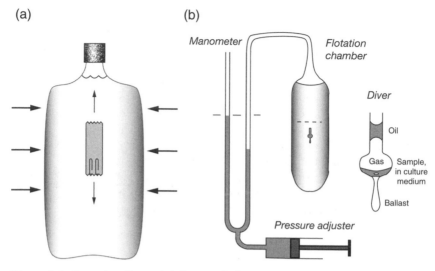

Figure 8.6. Cartesian divers. (a) One made from contemporary artifacts—a foil single-serving package of condiments ballasted just short of sinking with a few paper clips in a water-filled squeezable 2-liter plastic soft-drink bottle. (b) A diver used as a micro-respirometer, a simplified version of that described by Linderstrøm-Lang (1937). Today one would probably substitute an electronic manometer for the U-tube.

within the diver then allows calculation of the volume of gas consumed. With a manometer indicating pressure differences of about 20 pascals, and with a good way to adjust pressure, a volume change of 2,000,000 cubic micrometers (2×10^{-6} milliliters) can be measured. Zeuthen (1943) managed to increase sensitivity to 2,000 cubic micrometers and with the instrument measured the oxygen consumption of single amoebae.

Some years later, Scholander et al. (1952) improved the technique further, adding a reference diver for nulling the pressure and isolating the sample in a bubble whose volume could be measured with an ocular. This more complex version had yet greater sensitivity, about 200 cubic micrometers—the volume of a sphere about 7 micrometers across or of a typical animal cell—good enough to measure the metabolic rate of still smaller cells. Per Scholander, of course, was the great master of manometry, both macro and micro—recall his device for measuring the extreme negative pressures in the vessels of plants, described in the previous chapter. These divers have not so much been superceded as become victims of waning interest in the kinds of things they could measure.

GRAVITY VERSUS EVOLUTION

This extended consideration of the biology of gravity—three chapters in all, with numerous facets left untouched—prompts one final note. Once established, we put theories or laws or definitive equations to two fairly distinct uses. Sometimes we ask that they explain phenomena in the world around us; sometimes we ask that they predict some future state of that world or the outcome of some deliberate manipulation. To explain is not necessarily to predict, and predictions need not depend on intuitively satisfying explanations. Some theories do better at one task, some at the other.

Gravity, expressed as Newton's universal equation, does a splendid job of prediction. Our contemporary technology, especially in its larger manifestations, would be unthinkable without its reliable precision. But as an explanation, I think it serves us poorly. Taken at face value, it requires that every bit of matter in the universe have some sense of the existence of every other bit of matter. Modern physics (so I am reliably informed) does not rely on such a metaphysical assertion, but its alternative explanation lacks intuitively satisfying persuasiveness.

Evolution, as defined by the concept of natural selection, has the opposite virtue. It does a fine job of explaining both the large- and the small-scale phenomena of life, including many subtle and even counterintuitive observations. It does make predictions, but their precision never approaches that of Newton's simple and succinct equation. The real phenomena whose futures we want to predict involve too many players, too much contingency, too much amplification of insignificant perturbations. This contrast plays roles both in determining the relative status of biology among the sciences and in fueling the criticism of evolution from some quarters outside the sciences.

CHAPTER 9

Making and Maintaining Liquid Water

WE'RE ALL WET

Metabolically active organisms contain water in its liquid phase—I know of no exceptions. Life's domain consists of the places on the earth's surface where liquid water will persist—or places where we artificially ensure that condition. No single phase of a single compound so characterizes the conditions necessary for life. Yes, a few organisms tolerate a few ice crystals, usually if extracellular. And water vapor plays useful roles—in reducing evaporation, as a condensable resource, perhaps for producing the density variations that permit free convection. But liquid water is crucial to life as we know it. In a once well-known book, *The Fitness of the Environment*, Lawrence Henderson (1913) assigns a definitional rather than merely a facilitating role to the particular and peculiar physical and chemical properties of liquid water. The specifics of his argument, one might say, still hold water, even if their context now strikes the reader as offensively teleological and tautological and even as it fails to persuade—at least as contemporary biologists understand fitness.

Water abounds on earth, and most of that water remains in its liquid phase under conditions typical of its surface. But life perpetually pushes against environmental limits—in particular, against its abiotic barriers. Managing where water is minimally accessible can open otherwise unexploited regions or provide refuges, permanent or temporary, from biotic challenges. In all too many places, temperatures sometimes fall to levels at which water prefers its solid phase. Nor can life ignore its gaseous phase. While temperatures exceed water's normal boiling point in few habitable places, vaporization occurs in virtually all terrestrial habitats.

So familiar is that last point that a subtle peculiarity of our immediate world easily escapes notice. Enclose a dish of some volatile liquid in an air-filled container and keep the whole thing in a dark place at a constant temperature. Evaporation will proceed until the gas phase contains vapor at its saturation partial pressure, at whatever concentration corresponds to 100 percent relative humidity for that substance. But even over large bodies of freshwater, atmospheric air rarely reaches 100 percent humidity (or 97 percent, when equilibrated above seawater). Temperature variation, convection, and wind create innumerable opportunities for condensation. And thus some of us (still) hang out our washing even on overcast days, confident that it will dry in the air. Elsewhere equilibration can and does occur. For instance, the air within soil contains water at full saturation pressure except in the few top centimeters of the soil of deserts (Schmidt-Nielsen and Schmidt-Nielsen 1950).

Coping with Ice

Ice may be the most widespread toxic substance afflicting life. Not that one can't imagine positive roles for it. Even at constant temperature melting ice absorbs energy, so it could provide cooling elsewhere in a system—thus we cool beverages, freeze ice cream, and so forth. It can provide an osmotically inert and minimally volatile store of liquefiable water. Fractional crystallization as it forms can concentrate desirable solutes. Its lower-than-liquid density suits it as a flotation device. Adjusting the ratio of liquid to solid water in a system at the freezing point can stabilize temperature, permitting a kind of 0°C homeothermy. But as far as I know, all of these possibilities remain biologically hypothetical.

One can, though, point out a few instances where ice finds use. Ice has a low thermal conductivity so it sometimes provides insulation, especially in the flocculent form of snow, a cheap and disposable (but not easily portable) augmentation for fur or feathers. Water's high heat of fusion means that relative to deposited mass, frost formation will buffer a nighttime drop in plant temperature from radiation to the sky better than will dew. Just as melting ice absorbs energy, freezing water releases it. So a slight temperature increase will accompany ice formation within organisms. The reader might cast about for other possible cases—recognizing a possibility can provide a sieve, a search image, or a hypothesis.

Again, ice can't be regarded as healthy for organisms. The phase change from liquid to solid presents a far greater challenge for active organisms than does low temperature per se. Perhaps the worst aspect of the ice problem is that, unlike other environmental solids, it forms spontaneously from something that organisms absolutely require, at temperatures they often encounter. Pure water freezes at 0°C; a 1 molal (M) solution freezes at −1.86°C, and freezing point depression tracks solute concentration fairly linearly. Seawater of ordinary salinity, 35 parts per thousand, freezes at about the same temperature, −1.85°C. The bloods of teleost fishes freeze at warmer temperatures, −0.55° to −0.75°C, with values for marine fishes a bit lower than those for freshwater forms—but still above those of the surrounding water. The cellular fluid of young leaves freezes at about −0.56°C (Nobel 2005).

What might a creature do if exposed to temperatures below the freezing point of its internal fluids? Of the possibilities, avoidance must be the most straightforward, and it's not at all uncommon. Soils have low thermal conductivities, so where year-round mean temperatures exceed 0°C—and so are free of permafrost—burrowing or moving into caves provides refuge. In addition, while water may be prone to solidification, the solid phase has a lower density than that of the liquid. As a result, freshwater lakes stratify as they cool toward the freezing point, with ice forming on top only after the whole body of water has cooled below the temperature of maximum density, 4°C. Thereafter, contact with cold air above merely thickens the ice, and it does that only slowly due to ice's low thermal conductivity. Nor need life hold itself in abeyance below. The lack of wind-driven stirring and the diffusion barrier at the top suggest that oxygen should be in short supply. But fortunately ice transmits light, so photosynthetic flora can support an active community beneath the ice—short daylength may limit activity as much as low temperature. In some locations the oddly masochistic sport of fishing through holes in lake ice has become a major recreational activity.

A less-than-obvious way to avoid freezing consists of avoiding ice itself even while living at a temperature below one's own freezing point. The trick takes advantage, first, of water's propensity to supercool without (if internal) notable physiological effect, and then of the way it rises just prior to freezing, keeping the depths of a body of water reliably ice-free. Fish living deep down in some polar estuaries do just this trick. With no local ice to nucleate freezing (sometimes referred to as "inoculation"), they can spend entire winters at least slightly supercooled. Bringing one up to an icy surface or (in the laboratory) touching such a supercooled fish with a bit of ice produces an immediate, lethal wave of solidification (Scholander and Maggert 1971).

Supercooling turns out to be common; its practical lower limit, about −10°C, imposes its main drawback. Pure water, especially very small samples, can be supercooled much further, but organisms seem unable to achieve the requisite exclusion of ice-nucleating substances, inert particles, and microorganisms. Curiously, large animals can supercool less far than can small ones, whether they're compared intra- or interspecifically—apparently larger size comes with a higher likelihood of harboring ice-nucleating agents. Lee and Costanzo (1998) give a good review of cases and mechanisms.

Substances that lower the freezing point of the water in organisms—"cryoprotectants"—provide a common alternative (or sometimes an addition) to supercooling. We know of a variety of biologically employed compounds. Sugars and related compounds, including glycerol, glucose, sucrose, trehalose, and sorbitol, increase the osmolarity of solutions without changing their ionic compositions. These find widespread use in terrestrial plants and all kinds of terrestrial invertebrates and vertebrates. Overwintering insects may contain as much as 25 percent glycerol, which not only depresses their freezing points but permits supercooling well below those points—to almost $-50°C$ in a few species (Willmer et al. 2000). Polar fishes and various other systems use glycopeptides and proteins as antifreezes; these work in some way other than by altering the bulk colligative properties of body fluids (DeVries 1983; Lee and Costanzo 1998).

Such proteinaceous antifreezes, extracted from fish, have received attention recently as imposed cryoprotectants for living material such as sperm for insemination and organs for transplantation (Fletcher et al. 2001). As biosynthesized cheaply and abundantly by cooperative bacteria, they hold promise for the production of stable ice cream and other frozen desserts without either the usual fat or sugar—and thus for "diet" versions.

Many organisms eschew cryoprotectants and simply freeze—or not so simply, as they control both the sizes and sites of ice crystals. In all but a few instances, survival requires that solidification remain extracellular. The general lethality of intracellular ice has been recognized at least since a classic paper by Chambers and Hale (1932). But at least two very different kinds of animals, a nematode and an insect, tolerate intracellular ice, so it's not inevitably fatal—as noted by Wharton (2002), who did the work on nematodes. While we can expect other cases to be uncovered, they'll undoubtedly remain exceptional. Any case, though, challenges us to explain what makes some cells ice-tolerant.

Ensuring that ice will only form outside cells takes at least two devices. First, since water supercools readily, an organism has to provide something to nucleate extracellular freezing and to do so before body fluid becomes too metastable and solidifies too quickly. It's fairly clear that to be tolerable, freezing should be slow. So-called ice-nucleating proteins, highly hydrophilic, commonly serve this purpose. In a sense supercooling and extracellular freeze-toleration are antithetical—the greater the degree of supercooling, the more rapidly the succeeding wave of solidification will travel. Unsurprisingly, the two only rarely appear in the same places.

Second, an organism must ensure that externally initiated ice crystals do not penetrate its cells. Chambers and Hale (1932) looked quite specifically at the matter in onion epidermal cells, in amoebae, and in frog muscle fibers. In each case, cell membranes provided remarkably effective barriers to crystal propagation. Their manipulations leave no doubt that extracellular freezing and intracellular supercooling can occur simultaneously. Since ice easily penetrates most other biological structures, even macroscopic ones, that combination is remarkable. We should note, as well, the remarkable extent of that entirely extracellular solidification. The water in intertidal periwinkle snails may be as much as 80 percent frozen (Kanwisher 1955; Murphy 1983), some frogs survive freezing of up to 65 percent of their water (Storey and Storey 1992), and ice levels in a wide variety of reptiles can reach 50 percent (Storey 2006). Extracellular ice may occur in plants, but curiously they don't use either the antifreeze or the ice-nucleation proteins so widespread among animals (Guy 1990).

Keeping ice outside one's cells does not render it entirely benign. Besides being mechanically disruptive and an impediment to movement—sand in the gears, so to speak—it must present a severe physiological challenge to cells and tissues. Unless frozen very rapidly ("flash-frozen" in the food industry), water excludes salts and most other solutes as it solidifies. Thus, what liquid remains becomes ever more hypertonic. Hypertonic liquid outside cells draws water from inside. While that may lower the freezing point of intracellular liquid, it should affect a wide variety of cellular processes as well. Frozen snails may remain alive, but their metabolic rates drop to levels far below extrapolations from ordinary temperature curves (Kanwisher 1959). Plant cells lose turgor, often with macroscopically obvious damage to the supportive systems of their herbaceous parts or of whole plants (Sakai and Larcher 1987).

Freeze-Induced Outgassing

Besides making mechanical and osmotic mischief, freezing living material may create another problem, one that has drawn less attention. As the temperature of liquid water drops, more and more gas can dissolve in it. Solidification not only arrests the trend but radically reverses it. Far less air can remain dissolved in ice than in cold liquid water, at least a thousand times less according to Scholander et al. (1953). So water outgasses as it freezes, and ice cubes made from tap water in a household freezer come out cloudy. (Icicles, though, may be clear since they freeze outward rather than inward and thus exclude air.) Thus freezing also commonly involves the third state of matter. The greatest gas problem occurs in the most embolism-sensitive biological system, the xylem through which sap ascends in trees. Where winters are severe, most xylem vessels annually freeze and embolize; they must be refilled by positive pressure—generated by osmolyte release in the roots—each spring (Nobel 2005). The death of most broad leaves each fall may in fact represent adaptive recognition of the incipient failure of their water supply. Even the much narrower tracheids of evergreen conifers may suffer embolisms that must be repaired each spring (Sparks et al. 2001; Mayr et al. 2003).

Air occupies volume, so the density of frozen water in nature may be significantly lower than the published values for pure ice. Unsurprisingly, the expansion itself may damage tissue. Nor will the air work its way out by diffusion at any reasonable rate. The diffusion coefficients through ice for the gases making up air are lower even than the low coefficients for diffusion through most other solids, something also pointed out by Scholander et al. (1953).

Nor will the outgassing be limited to the extracellular areas where ice crystals form. Adding salt to a solution decreases the solubility in it of most nonpolar substances, the so-called salting-out effect. (Edsall and Wyman 1958 give a good account of the phenomenon from a physiological perspective.) A commonly cited formula for the effect equates the logarithm of the ratio of solubilities (S) in adulterated and pure solvent with the product of minus one, the "salting constant" (K_s), and the molality of the solution (C_s):

$$\log\left(\frac{S_{solution}}{S_{solvent}}\right) = -K_s C_s \qquad (9.1)$$

Salting constants, in practice obtained from tabulations, range from about 0.1 to 0.4 liters per mole. The formula applies only to dilute solutions, so since the effective molality of what remains after a cell loses much of its water is both high and far from certain, it's only a rough guide. But it's generally consistent with the observation that about 20 percent less air dissolves in seawater than in freshwater (Krogh 1941). As mentioned, seawater's freezing point depression of 1.85°C equals that of a 1 molal solution. A salting constant of 0.1 liters per mole produces about this value, fairly close to the value of 0.13 liters per mole cited by Edsall and Wyman (1958) for oxygen in a solution of NaCl. Seawater, of course, should be less salty than what remains after half of the water in an organism has become ice.

So salting out of dissolved gas may be substantial in both volume and effect. (However, nucleation of bubbles rather than salting out explains the sudden outgassing when one adds a pinch of salt to a carbonated beverage.) An indication of the effect may be from the reports of Kruur et al. (1985) and Lipp et al. (1987) of better recovery of mammalian cells from repeated cycles of freezing when suspended in degassed solutions. At least gas released inside cells might in part offset the concomitant osmotically induced volumetric shrinkage.

A tissue may face yet another consequence of the exclusion of air from extracellular ice and of salting out. Surface tension as well as atmospheric pressure squeezes bubbles. As a result, the pressure within a tiny bubble of gas in a liquid depends on its size, with the effective squeeze of surface tension varying with its radius of curvature. The last chapter invoked the phenomenon when looking at the stability of gas bubbles in water and at the spacing of hairs in the plastrons of aquatic arthropods. To repeat, then, for a spherical shell with a radius of curvature of r and a corresponding diameter of d, the pressure developed by surface tension (γ) is

$$\Delta p = \frac{2\gamma}{r} = \frac{4\gamma}{d}. \tag{9.2}$$

Equation (9.2) implies that coalescence of a group of gas bubbles will produce a bubble of greater volume than the combined volumes of its contributors—simply by being bigger, it will be squeezed less. How much so? Since the product of pressure and volume must be maintained in the coalescence, the solution (mathematical, not physical) is simple enough. Volume (V) varies with the cube of radius, and pressure varies inversely with radius (as above), so, for coalescence of n bubbles, each of r_i into a final one of r_f,

$$r_f^2 = nr_i^2. \tag{9.3a}$$

Recognizing that the volume of the product represents n bubbles,

$$V_f = n^{3/2}V_i. \tag{9.3b}$$

Combine 100 bubbles and the resulting one will have 10 times the sum of their original volumes.

The specific value of surface tension turns out to be irrelevant—while it drives the phenomenon, any amount will do. The formula, though, ignores the atmospheric contribution to pressure so it's limited to cases in which the latter can be neglected; on earth that means infinitesimally small gas bubbles whose internal pressure comes overwhelmingly from surface tension.

Accounting as well for the squeeze of atmospheric pressure puts the value of surface tension into play, and it puts absolute size back into the picture:

$$r_f^3\left(A + \frac{2\gamma}{r_f}\right) = nr_i^3\left(A + \frac{2\gamma}{r_i}\right), \tag{9.4}$$

where A represents atmospheric pressure. For 100 bubbles, as earlier, the volumetric expansion factor expansion no longer reaches 10; for 1000

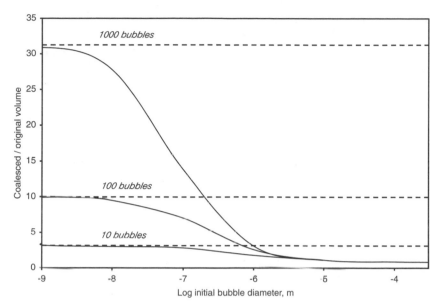

Figure 9.1. The factor by which equal-sized bubbles of various sizes increase in volume when they coalesce into a single, larger bubble.

bubbles it no longer reaches 31.6. Figure 9.1 shows what equation (9.4) says for the coalescence of 10, 100, and 1000 equal-sized bubbles as a function of bubble diameter. For an initial diameter of 0.1 micrometers, the expansion when 1000 bubbles coalesce is fourteen-fold; for an initial diameter of 1 micrometers, it's 3.3-fold; for an initial diameter of 10 micrometers, it's only 1.3-fold. For an initial bubble at the threshold of direct visibility, 100 micrometers, the expansion factor drops to a trivial 1.03-fold.

In short, at ordinary atmospheric pressure, this extra expansion during coalescence from the reduced effect of surface tension, minimally relevant to everyday foams, might be distinctly nontrivial at a cellular scale.

But is outgassing, the parent phenomenon, significant? The existence of winter embolisms in the xylem of trees, mentioned earlier, is well established. A number of observations and passing comments suggest other cases that might bear investigation. The clearest come from Scholander et al. (1953), with such statements as, "A principal difficulty in experimentally freezing animals in water is that gases are trapped in the ice and form bubbles which rupture the tissue." It reports that upon thawing, naturally frozen copepods sometimes develop fatal internal gas bubbles.

Pre-freezing exposure to seawater of higher salinity and lower oxygen tension—both of which mean less dissolved gas—improved survival of intertidal mollusks beyond what would be expected from their increased tissue osmolarity (Murphy 1983). Again, Kruur et al. (1985) and Lipp et al. (1987) found that mammalian cells recovered better from repeated freeze-thaw cycles if the cycling was done in degassed solutions.

Evidence of troublesome coalescence during thawing is at this point merely suggestive. Slow thawing, which would permit better redissolution of gases, usually gives better survival than does rapid thawing. Murphy (1983) mentions that intertidal snails do better if kept cold following thawing, which also would permit better redissolution.

MINIMIZING EVAPORATION

Normally water won't solidify unless the temperature drops below the freezing point of its particular solution. But it can vaporize at any temperature. Even ice can transform into the gaseous phase. Although slow, such sublimation often adversely affects the appearance and subsequent palatability of poorly packaged frozen food. Few terrestrial habitats exposed to the atmosphere can escape some evaporation or sublimation of water. Nor can organisms that depend directly on atmospheric gases easily seal themselves off from evaporation—whether they breathe gaseous oxygen for respiration or absorb carbon dioxide for photosynthesis.

In biological terms, what differentiates evaporation from solidification is that evaporation has a demonstrably positive role. Water's heat of vaporization is especially high, 2.44 megajoules per kilogram at 25°C. That's greater, incidentally, than the commonly quoted 2.26 megajoules per kilogram at 100°C because of the additional hydrogen bonding at the lower temperature. For comparison, the heat of vaporization of propanol, with about the same boiling point, is 0.76 megajoules per kilogram, over three times lower. That makes evaporation of water an especially good way to absorb heat for, say, disposing of excess heat from atmospheric or solar input or from metabolic inefficiency. One can tolerate environmental temperatures well above that of one's body (as in a sauna), as long as a low enough humidity permits sufficient evaporative cooling. What limits the use of evaporative cooling is the supply of water. While the oxidation of food yields metabolic water, the subsequent vaporization of that water absorbs much less heat than what's produced in that metabolic process. So metabolic water cannot absorb sufficient heat to have much effect on body temperature, and cooling by evaporating water requires some more copious source.

Only medium and large terrestrial animals, mainly mammals and a few large birds, control body temperature through evaporation, that is, by panting or sweating. That's almost certainly because for the process to be practical, an animal's surface-to-volume ratio cannot be too high. And those warm, open habitats where cooling may be critical tend to be ones with limited supplies of water.

More problematic are broad leaves, many of which get quite warm when in sunlight and which evaporate ("transpire") water at substantial rates. That evaporation must produce significant cooling, so its functional significance is undeniable, and energy budget accountings rarely ignore it. By contrast, its adaptive significance has long been controversial. Traditionally, evaporative water loss has been regarded as an unavoidable evil, an unfortunate consequence of keeping leaves sufficiently permeable for inward diffusion of photosynthetically critical carbon dioxide during daytime.

At least two facts argue against the generality of the traditional view. For one thing, water use efficiency, the rate of CO_2 fixation relative to water loss, varies considerably, almost tenfold from plant to plant, even among those for which water isn't freely available. If there's an irreduceable minimum cost in water for obtaining CO_2, we'd expect plants that are routinely drought-challenged to hug that upper limit on water use efficiency. For another, many xerophytic (dry habitat) plants, ones representing several evolutionary lines, simply close their stomata during the daytime so they neither evaporate much liquid water nor absorb much CO_2. Instead they open them and absorb their CO_2 at night, when the air is at or near the dew point, and store it as organic acids for decarboxylation during the day. They thereby raise water use efficiency as much as ten times further. Even so, one should note, a lot of water emerges from a plant relative to the amount of carbon dioxide fed into the photosynthetic process—or relative to photosynthetically fixed water. For mass of water relative to mass of CO_2 (the inverse of water use efficiency), ratios run between 25 and 1000 (Nobel 2005), vastly higher than the chemically minimal 0.41. Not only does water diffuse more readily than CO_2, but the past activities of the plants themselves have left the atmosphere with very little CO_2—now a little over 0.03 percent but even less just a few years ago. And few plants ever enjoy the luxury of photosynthesizing at 100 percent relative humidity.

As we terrestrial animals seek oxygen, we face an analogous balancing act to that of plants looking for carbon dioxide—an exchange surface that takes in oxygen will usually lose water. Still, our situation isn't as dire. The mass of oxygen absorbed to mass of water lost is 1.6 for a human, over an order of magnitude better than even the most efficient plants.

Moreover, our metabolism yields rather than consumes water. For each 2.1 mass units of oxygen used, fully catabolizing pure starch yields a unit of water. So we can compensate for most—but not all—of our respiratory water loss from that metabolic yield. Moreover, our situation is more favorable in another way. Oxygen may have a slightly higher diffusion coefficient than does CO_2; more importantly, its atmospheric availability is far greater, hundreds of times so in terms either of relative mass or of molecular concentration.

Nonetheless, getting oxygen still requires that precious liquid water disappear into thin air. We might envy the access of well-rooted plants to the interstitial liquid water in the soil. Alternatively, that access might be seen as a severe constraint on mobility, one we rarely consider. In other words, rootedness may be critical to the ability of plants to transcend what, by animal standards, are extremely low water use efficiencies.

We humans, though, win no prizes among animals for the efficiency with which we process water. Our kidneys cannot, for instance, produce urine even as salty as ocean water, and thus we dehydrate if we drink seawater. Besides having more effective excretory systems, many mammals and birds that are either small or lack reliable access to water have another way to minimize loss of liquid water. Chapter 5 described the nasal countercurrent exchanger, first recognized by Jackson and Schmidt Nielsen (1964), as a device for heat conservation. But it simultanously serves as a water-conserver. Inhaling cold, dry air through intricate nasal passages causes evaporation of moisture from the walls of those warm, wet passages, so air arrives at the lungs nearly saturated at body temperature. At the same time, the exchanger cools the nasal passages, so in the subsequent exhalation moisture condenses as heat is removed. And so air, departing only a little above atmospheric temperature, is saturated with water at that temperature rather than at the body's.

The arrangement conserves significant amounts of water. For a kangaroo rat, respiratory water loss over oxygen use is 0.41 rather than our 0.64 or a laboratory rat's 0.72. Significantly, 0.41 is less than the metabolic yield from metabolizing starch, 0.47 (the reciprocal of the 2.1 above). As a result, kangaroo rats and some other desert rodents need drink no liquid water at all—unless they eat a high-protein diet and need water to dispose of the excess nitrogen.

A greater difference between core body temperature and air temperature ought to increase the degree of conservation. Although no specific investigation appears to have been done, musk oxen (*Ovibos moschatus*), once common in the New World Arctic, may make exceptionally effective use of this temporal countercurrent arrangement. Their daily water turnover—loss relative to body mass—averages only 3.5 percent, only a little more than half of a camel's 6.1 percent and far less than the

13 to 20 percent of cattle and water buffalo (Prosser 1973); winter turnover is at least five times less than the summer rate (Klein 2001). Concomitantly, musk oxen have unusually wide nasal openings, which should slow airflow, and elaborate nasal turbinates, providing more surface for condensation. Plate 9.1 provides a comparison with a domestic cow.

If musk oxen do use the device, one wonders about its primary adaptive advantage. The extreme coldness of the air they breathe during the long winter means that without a trapping system air arriving at their lungs will be almost devoid of water vapor and respiratory water loss will be high. They may be surrounded by snow, but liquefying it takes energy—6 to 14 percent of their overall expenditure (Soppela et al. 1992). Still, heat trapping may matter as much as minimization of water loss, contrasting sharply with camels, for whom reduction in heat dissipation must be a drawback.

Extracting Liquid Water from Vapor

Since all active organisms require liquid water, and since gaining either oxygen or carbon dioxide from the atmosphere inevitably entails loss of water, air-processing terrestrial organisms (except the few that can manage with metabolic water) must be able to acquire it. In no habitat does the atmosphere entirely lack water vapor. So an organism just needs to condense liquid from the limitless surrounding supply. But only a few, mainly arthropods and vascular plants, do just that—for some reason (or reasons) the task is daunting.

We can recognize several possible mechanisms for condensation: (1) exposure of a solution concentrated enough and thus with a sufficiently low vapor pressure to draw water from the local air; (2) exposure of a hygroscopic surface followed by extraction of water from that surface; (3) exposure of a negatively curved air-water interface (an "antibubble") with a very small radius of curvature; and (4) exposure of a surface with a temperature below the dew point of the air around it.

Organisms commonly transport ions actively and concentrate nonionic osmolytes. But making solutions that condense water vapor faces an unfavorable relationship between concentration and the relative humidity of the air at equilibrium, at least with ordinary substances as solutes. For instance, a 3 molal solution of NaCl (16 percent by weight) will only condense water vapor from an atmosphere of at least 90 percent relative humidity. For condensation, a 4 molal-thick solution requires a humidity above 86 percent. Further complicating things, few if any organisms can actively transport water, as opposed to ions or other solutes.

So absorption of condensed water has to be done indirectly by such strata-gems as pinocytosis of solutions followed by extraction and ejection of solutes. Both the first, simple colligative condensation, and the second of our mechanisms, use of hygroscopic substances, face the problem.

A variety of arthropods do manage to condense water from substan-tially subsaturated atmospheres, and they depend—although specific de-tails remain obscure—on some mix of colligative and hygroscopic devices. Some do so from remarkably dry air. Several kinds of fleas, booklice, and mites can extract liquid water at relative humidities below 80 percent, while some lice, a few silverfish, and a beetle larva can deal with humidi-ties below 50 percent. In all cases investigated, absorption occurs at spe-cialized surfaces, rectal, oral, or Malpighian (insect renal) (Willmer et al. 2000). Arthropods of deserts such as the Namib of Southwest Africa take advantage of the relatively moist air coming off the adjacent ocean; some use condensation of vapor, others collect liquid water from fogs (Hadley 1994). One case of condensation from an unsaturated atmosphere has been described in plants, a succulent shrub in the Atacama Desert of Chile (Mooney et al. 1980), another desert east of an ocean.

Evaporating water takes a supply of energy equal to its heat of vapor-ization, hence its utility as a way to offload excess thermal energy. Con-densation, its opposite, must yield energy, warming the condensate and anything in thermal contact with it. No animals have been reported as using that thermal energy, which isn't unsurprising, given the low surface-to-volume ratios of animals compared with, say, leafy plants—for which a thermal role for dew formation was suggested earlier. Worse, water must be moved from the solute rich sludge in which it may have con-densed to some higher concentration in the body fluids, and those severe osmotic gradients will impose a significant price—although, even so, the cost of condensation relative to an arthropod's other activities turns out to be fairly low (Hadley 1994).

For the third of our possible mechanisms, use of locally lowered pres-sure, we know no specific case; and it's probably unlikely. Willmer et al. (2000) mention muscular pumping to produce pressure cycles, but only as a hypothetical. One might imagine using the opposite of the surface-tension induced compression of bubbles. As noted when talking about plastrons in the last chapter, pressures can be reduced by air-water inter-faces with negative curvature, ones in which reduction of the area of the interface would decrease rather than increase pressure. For a plastron, pressure was lower on the gaseous side. In the present terrestrial rather than aquatic situation, hydrophilic rather than hydrophobic pegs or water-filled grooves between ridges might cover a surface and lower the pressure on the liquid side. Atmospheric water vapor would then condense into the low-pressure liquid. But the scheme will not work. As long as the surface water remains liquid, its concentration will far exceed that in the adjacent atmosphere at any humidity. If it did work, the highly negative pressures in their xylem would permit the leaves of tall trees to extract atmospheric water—a possibility they'd certainly not overlook.

For the fourth mechanism, condensation at a locally subatmospheric temperature, we have no shortage of cases, at least among plants. We animals can achieve subambient body temperatures, but we inevitably do so by evaporative cooling, entirely inappropriate if gaining liquid water is the goal. We might use radiation to the night sky if we can meet a few conditions. The humidity (or the dew point temperature) must be high, the sky must be nearly or entirely cloudless, the atmosphere near the ground must not be obscured by smoke or fog, the surface must be arranged to limit convective reheating by the surrounding air, and air movement must be minimal, for the same reason. We're of course defining the conditions for forming dew, conditions especially well met by low, broad-leaved, or succulent vegetation in open areas, especially where trees surround but do not cover areas of vegetation.

As Nobel (2005) points out, the requisite high humidity near the ground may come about as much from distillation from the soil as from the atmosphere itself. Real wind can dry a surface as rapidly as condensation occurs, keeping ground-level humidity low. Still, gaining liquid water from an atmosphere kept humid by soil distillation takes at least some slight air movement—in truly still air, diffusion of water vapor will be insufficient to offset very local depletion by condensation. As it happens, air movement should always be at least adequate. Substantial quantities of water may be condensed, up to 0.5 millimeters per night or 30 millimeters annually. Environmental structures occasionally provide the effective surfaces—Hadley (1994) mentions rocks that drip water from condensation after radiative cooling. Whether on vegetation or other structures, condensed water may be imbibed by desert arthropods (Edney 1977). In effect they outsource radiative cooling and condensation where their own small sizes and bulky shapes preclude the process.

We may at least occasionally condense moisture as a source of drinking water. Jackson and van Bavel (1965) described a lightweight, inexpensive survival device for use in arid places, shown in figure 9.2, with which they collected up to 2 liters of water per day. It consisted of nothing more than a piece of reasonably hydrophilic, clear plastic film about 2 meters in diameter and a wide-mouth container to catch the condensate that drips down the film—and a shovel to dig a pit. Sunlight passing through the film heats the soil (and any moisture-containing plant material that may be added to the pit), evaporating water that then condenses

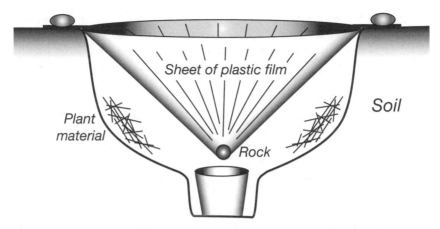

Figure 9.2. The solar still of Jackson and van Bavel (1965). The pit is about 1 meter in diameter and 0.5 meters deep.

on the film. A similar device, designed for use on life rafts, plays a central role in a popular novel (Martel 2001).

CONDENSATION BY LOCATION CYCLING

Dropping body temperature below the local dew point by radiative cooling to condense physiologically significant amounts of water may be impractical for animals. Unlike plants, they have another option. An animal might shift back and forth between a hot and a cold location, chilling enough in the cold site so it can condense water on itself after moving to the warm one. The frequency of shifting will be in part set by its size since it has to stay in the cold site long enough to cool down. It should then stay at the warm site long enough to condense water, although water gain will drop as it approaches local temperature, and retreating to the cold site might give more time for internalization of the condensate.

One thinks immediately about cycles of immersion in some cold ocean alternating with emersion. But that would be of use only to the relatively small numbers of animals (such as ourselves) that cannot obtain body water from the ocean or perhaps to the inhabitants of the edges of hypersaline lakes. Cycling between two terrestrial sites seems better, as long as the sites are sufficiently close to permit quick and cheap repeated shifting. As far as I know, no specific case has been reported; if cases exist, they'll probably be uncommon. Still, I believe I've identified one, which I will describe below, prefaced by the candid admission of a serious gap in the chain of evidence.

Turret spiders (genus *Geolycosa*), relatively stout wolf spiders (plate 9.2 left), live mainly in sandy soil amid low, sparse vegetation or among coastal sand dunes (Wallace 1942). Each spider constructs an almost perfectly vertical burrow about a fourth to three-fourths of a meter deep and about 10 millimeters in diameter. It keeps the burrow from collapsing with silk lacing near the top where the soil is friable. It rests at the bottom during the day and straddles the top at night, then feeding on passing arthropods. The common name refers to the inevitable turret or crater at the top of the burrow, a construction of silk-stabilized sand and bits of vegetation that extends the burrow between a few millimeters and a centimeter above the ground (plate 9.2 upper right). The turret has been presumed a barrier to flooding, for which it looks too flimsy and porous, or as a look-out perch, which isn't borne out by my observations that spiders straddle their turrets.

These spiders live in well-drained habitats that lack standing water for long periods. But the soils around their burrows contain substantial amounts of interstitial water with, as a result, very high interstitial humidity—as noted earlier, as do soils beneath the immediate surface, even in deserts. While some spiders can extract interstitial water (Parry 1954), *Geolycosa* have not been reported to do so. Nor can they directly extract water from the air, even at 98 percent humidity (Humphreys 1975). In the warmer months, sunlight warms the surface well above the general air temperature, often to where walking barefoot is painful. So during the day a burrow develops a severe temperature gradient, warm at the top and (relatively) cool deeper down, as in figure 9.3. The consequent stratification (an "inversion" in the meteorological sense) ought to minimize internal mixing.

Perhaps a spider might obtain water by moving up and down within its burrow, climbing up near the top with a cool body, condensing water, and then retreating to the bottom to absorb the water and cool again. But no condensation was evident when slightly desiccated spiders were cycled between the conditions of temperature and humidity of the upper and lower portions of burrows in the laboratory. When the warmer air was made to flow slowly over the animals, though, droplets appeared almost immediately (plate 9.2 lower right). When the spiders were again in cooler air, they groomed themselves, the droplets disappeared, and they gained weight.

The necessity for some flow of air fits nicely with consideration of the Péclet number, speed (v) times length (l) over diffusion coefficient (D) (recall chapter 1). For $l=0.01$ meters and $D=2.7\times10^{-5}$ square meters per second (water vapor at 40°C), $Pé=1$ will occur at a speed of 2.7 millimeters

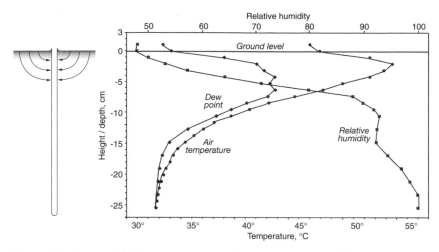

Figure 9.3. A typical mid-summer record of temperature, humidity, and dewpoint as a function of depth in a burrow such that of plate 9.2. In this particular burrow the maximum difference between bottom temperature and dewpoint is 11.1°C.

per second. Although very slow, that speed will exceed that of spontaneous air motion in a small, closed container—or a severely stratified burrow. At speeds significantly lower, the rate of condensation will be limited by diffusion. Since even warm and moist air still contains relatively little water, a spider will warm up before much condensation could occur.

That need for airflow may explain the role of the turret. Airflow across it will draw air up and out of the upper reaches of the burrow much more effectively than would flow across a flush opening (Vogel 1976), at least if the soil around it is sufficiently porous. The adequate porosity of the soil around these burrows was confirmed by measuring the pressure reduction needed to draw air out of burrows together with wind-induced pressure reduction in a model burrow. Enough air should indeed pass through the moist sandy soil and become humidified to maintain a supply of water vapor for condensation. As in figure 9.3, the dew point in the upper portion of the burrow will be well above the temperature near the bottom. While such induced airflow will be limited to the upper portion of the burrow, it would in any case be unnecessary further down.

About that logical gap. Despite several attempts, I never managed to observe spiders moving up and down during a dry spell in the summer.

Spiders are strongly vibration-sensitive. These particular ones were, in addition, all too quick to abandon burrows that suffered any kind of intrusion, even that of a tiny bead dangling from a thread. So the scenario remains conjectural until someone else pursues it. McQueen and Culik (1981) found that *Geolycosa* will go up and down around twenty times per hour, but in relatively disturbed burrows and under conditions of limited present relevance. Still, the rates they report are about right for getting water with the present mechanism. An object of about the thermal mass of a spider, when subjected to a 20° temperature increase, warms by 10° in a little under 3 minutes. A longer stay would warm a spider above the local dew point, and it would then evaporate rather than condense water. In fact, 3 minutes will overestimate the time available for condensation, since the process itself generates heat. So high is the heat of vaporization of water that condensation equal to 1 percent of body mass will raise body temperature by almost 6°C. Exposures of about a minute seems reasonable, with condensation limited to less than 1 percent per cycle and with ample intervals for internalizing the water at the bottom.

On the one hand, I think it more likely than not that these spiders do obtain liquid water by location cycling; on the other hand, the scenario illustrates the demanding requirements for exploiting the device. Just this point was made by Lasiewski and Bartholomew (1969), reporting on work with models, (their italics): "*The temperature and humidity conditions in these experiments were chosen to demonstrate that condensation could occur in nature. We do not necessarily mean to imply that it does occur.*" Still, many small desert animals build burrows in which they spend their daytimes (Wilmer et al. 2000), so we've ample candidates for its use.

That condensation depends on air movement raises one further point. Measurements of water extraction from nonsaturated atmospheres usually involve putting small arthropods in closed containers at a fixed humidity (the latter maintained by an appropriate salt solution elsewhere in the containers). If the container is too small or the temperature too close to constant for appreciable convection, then condensation will be abnormally slow. So figures for minimum usable humidity may be overestimates, and few reports give sufficient experimental detail from which to judge possible error—some animals may be doing even better than reported.

CRYPTOBIOSIS—LIFE WITHOUT WATER

For most organisms, drying below some critical water content means death. A taxonomically diverse minority endure the loss of virtually their entire content of unbound water—they appear to live without water. But we might put "live" in quotes, since they retain only one functional attribute of living things, that of not being irreversibly dead. In almost every case the addition of water restores normal activity after only a short period of physiological (and sometimes morphological) restoration. The state of suspended animation has been termed "cryptobiosis," with this specific version sometimes labeled "anhydrobiosis." For some organisms, cryptobiosis provides a way to wait out unfavorable conditions; for others it's a normal part of the way they disperse. In the cryptobiotic state organisms survive, not just their ordinary vicissitudes, but ones they would not normally encounter. For instance, drying of the habitat may induce the transition; but once cryptobiotic, the organisms can withstand moderately high or extremely low temperatures, high or low pressure, even extraterrestrial vacuum. Whatever the role of cryptobiosis, its consistent element remains that exclusion of liquid water.

The commonest example of a cryptobiotic creature is the brine shrimp, *Artemia salina*, normally an inhabitant of transient saline pools in deserts and sold as food for small fish. One can purchase jars of what look like coarsely ground grain. Add a bit to water and in a short time crustacean larvae swim off; the product has an almost indefinite shelf life. Not only some crustaceans, but many rotifers, nematodes, tardigrades, and collembola can enter cryptobiosis, as can vascular plants (as seeds), spores of many groups, and bacteria. We preserve many microorganisms by inducing cryptobiosis via lyophilization, that is, by freeze-drying them in a vacuum—keeping a type collection of such microorganisms takes far less upkeep than maintaining, say, stocks of mutant fruit flies.

The nonability of nonadapted multicellular organisms to enter cryptobiosis points to specific, requisite functional alterations. Use of cryoprotectants is common, most often the same ones, such as glycerol and trehalose, that permit nonfatal freezing. Prior to entering the state, surface area may be reduced by withdrawing appendages and rounding up and adding extra surface material may reduce permeability, the latter most likely to slow water loss during initiation (Wilmer et al. 2000). How long can organisms remain cryptobiotic? Wharton (2002) cites a claim of up to 250 million years for bacteria, but with a careful cautionary comment concerning the conundrum of controlling contemporary contamination. On reliable evidence, some invertebrates can apparently last over a century and some seeds more than a millennium.

Other devices to hold life's functions in abeyance such as aestivation and diapause represent less extreme shutdowns and require less extreme (if any) exclusion of liquid water. Cryptobiosis shows that while liquid water may be necessary for life, the absence of such water need not be lethal. In addition, it may indicate that nothing approaching suspended animation is compatible with the presence of liquid water. Put another way, life and death may be mutually exclusive, but together they aren't fully inclusive of possible states.

CHAPTER 10

Pumping Fluids through Conduits

PRESSURING THE PIPER TO PLAY

In the first chapter, I suggested that, because diffusion is ineffective over all but minute distances, an organism larger than a typical cell must move fluid to move material. By whatever name, internal bulk fluid movement absorbs energy as a consequence of that universal fluid property, viscosity. Supplying that energy requires some kind of pump. While the pump may also accelerate fluid or lift it against gravity, neither of these roles have quite the same inevitability. Not that the cost of pumping has to be metabolic. Sometimes an external agency can be coopted to do the work.

The diversity of circumstances under which organisms pump fluids, the phylogenetic diversity of the organisms, and the structural diversity of the pumps themselves—all stand in the way of treating biological pumps as a single class of functional devices. Separate books, or at least separate chapters, deal with the ascent of sap in a tree, the suction of blood by a mosquito, and the suspension feeding of a clam. Here I want to explore generalizations that might emerge from focusing the pumps themselves with all their functional, phylogenetic, and structural diversity just as context.

Pumps have also been ubiquitous in human technology since the first fields were irrigated with water that gravity alone could not supply, since water was first hoisted from lake, river, or well. Where industrial products have yet to reach agrarian cultures, a remarkably wide range of simply constructed but effective devices remain in use. Among our industrial products, only electric motors may exceed them in range of sizes, applications, and designs. When choosing among their diversity of pumps, engineers worry about such things as power expenditure, operating conditions, and efficiency. We biologists have only occasionally done the same in analyzing and classifying nature's pumps, and we ought to take full advantage of all that technological attention.

[In part, this chapter takes the same viewpoint as did a predecessor (Vogel 1995a), one that a reviewer rightly noted didn't turn out quite as satisfactorily as the author had hoped. I'll, of course, have to reiterate some of the points of that paper.]

The Relevant Variables

An insect, most famously an ant, can lift many times its own body weight—but it cannot lift it far. Similarly, a tree can draw sap upward with pressures of tens of atmospheres, millions of pascals—but it does so very slowly. We may be overly impressed by the spectacularly high forces and pressures that organisms can produce and insufficiently mindful of constraints on distances and volume flows. Whether lifting weights or forcing fluids through pipes, doing work involves three variables, and these operate over a wide range of combinations. Force, distance, and power define a lifting task, with power the product of force and rate of change of distance. Similarly, pressure, volume flow, and power define the task of a pump, with the last again the product of the first two. A bivalve mollusk can pump its own volume of water in, across its gills, and out again every few seconds, but it does so against only a few pascals of pressure—its volume flow should impress us as much as the pressure generated by a sap-lifting tree.

The particular graphic representation of pump performance shown in figure 10.1 has enjoyed long usage in engineering. With only a slight modification (incorporated here, as will be noted shortly) it should work as well for biological pumps. In practice we might be unable to measure performance of our pumps quite as far from normal operating conditions, but if need be we can extrapolate the end points of their basic operating lines.

The maximum pressure most pumps can produce occurs at zero volume flow, while maximal volume flows happen when the pressure they produce (or opposing pressure) is zero. So a curved line from one axis to the other, the "pump capability" line, along with the axes themselves, defines a potential area of operation—possible combinations of pressure (Δp) and volume flow (Q). In an actual application, operating conditions are constrained by the resistance of the load, defined by how much volume flow corresponds to each value of pumping pressure. For most technological pumps, the pressure needed varies with the square of the required volume flow. Thus, the "operating line" forms a parabolic curve extending upward from the origin. For virtually all biological pumps, Δp will vary almost directly with Q, so the line from the origin will be linear rather than parabolic. In either case, the intersection of this operating line with the pump capability line marks the maximum output of a particular pump in a particular application.

Size underlies that difference between technological and biological pumps. For the relatively small sizes and low speeds of ours, most flows will be laminar. Thus, the Hagen-Poiseuille equation or something similar

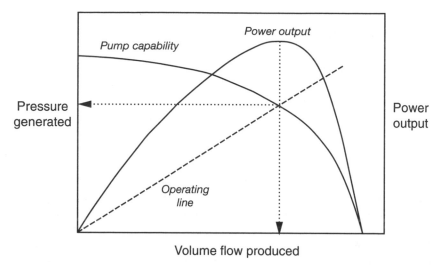

Figure 10.1. A pump performance graph. Most pumps yield plots similar to this one; they differ mainly in the scaling of the axes. The axis-to-axis outer curve ("pump capability") marks the limits of the pump under any operating condition. The dashed line corresponds to the operating condition set by the particular load imposed on this pump, while the dotted lines give the pressure and volume flow of that particular load. This example shows an optimally efficient combination of pump and application—the intersection of operating and pump capability lines lies just beneath the point of maximum power output.

applies; it defines a constant resistance ($\Delta p/Q$) analogous to that of Ohm's law for electrical resistance.

The power (P) a pump puts out equals the product of its volume flow and the pressure boost it gives the fluid passing through it—$Q\Delta p$. In practice the graph indicates power on a second ordinate. Its curve begins and ends at the abscissa since zero values of either pressure increase or volume flow mean zero power output.

For an ideal match of pump to task, the peak of the power output curve should lie just above (or below, since the ordinates have different scales) the point of intersection of the operating line and the pump capability line. The graph thus gives a valuable view of that match. If peak power output occurs well to the left of the intersection, the pump produces too much pressure and too little volume than would be best. It can't reach peak power output, in an application with that operating line. Conversely, if peak power output occurs to the right of that intersection, the pump produces too much volume flow and too little pressure; again, its maximum power output won't be realized. This second mismatch can have a particularly serious consequence if, as when lifting from a well or from ground to top of tree, producing any useful output takes some minimum (here gravitationally determined) pressure—a pressure threshold must be exceeded to get any flow.

Specifying the power output of a pump, or even power output plus energy conversion efficiency, may not tell us whether a particular pump gives proper service for a particular task. Pumps vary widely in the mixes of pressures and volume flows they can produce—in the specific shapes and positions of their operating lines. While all graphs of the sort shown in figure 10.1 may look similar, the scales on their axes will be anything but.

A FUNCTIONAL CLASSIFICATION OF PUMPS

The literature on pumps for technological uses (for instance, Karassik et al. 2000) recognizes diverse implementations of each of two general categories. Only a few devices fail to fit comfortably into one category or the other. (The specific names of each, though, differ a bit from source to source, causing awkwardness for online searching or using indices.) A few words about the devices in each category might stimulate recognition of biological equivalents—beyond the obvious ones.

Those in one category are most often called "displacement" or "positive displacement" pumps. Typically fluid is drawn into a chamber and then persuaded (as by reducing the chamber's size) to leave by a different route. Most familiar are ones with pistons that move back and forth in chambers with valves to ensure unidirectional, if pulsatile, flow—for instance the ones with which we hand-inflate pneumatic tires. A less common version, the diaphragm pump, changes chamber volume with periodic pushes against a flexible element that forms one of the chamber's walls; it exchanges the problem of a closely fitted piston for that of a nonrigid element. Other displacement pumps work by translocating the functional chamber itself. In the commonest versions—gear, screw, vane, and lobe pumps—multiple moving components carry fluid along as they themselves move. In another displacement pump, the so-called air-lift (or gas-lift) pump, bubbles of gas rising through a narrow vertical tube of liquid carry slugs of liquid upward between adjacent pairs of bubbles. Similar to these latter types, and of especial biological relevance, is the peristaltic pump, with its traveling constrictions of flexible tubes. It eliminates contact between fluid and pump housing and tolerates flows of variable viscosity and of fluids with suspended solids. But the technological versions perform inefficiently and are not particularly reliable, so they remain uncommon. The flows produced by displacement pumps range from nearly steady to severely pulsatile unless paired with some external buffer.

Those in the other category are called "dynamic," "fluid dynamic," or "rotodynamic" pumps; this last name recognizes their ordinarily rotational operation. All depend on fluid dynamics rather than fluid statics. The commonest drive fluids with axial or centrifugal fans and are most familiar as propellers and squirrel-cage air blowers. Another type is the jet pump, in which fluid gets pushed through a channel or duct by an axial squirt through a jetting orifice or "eductor" into its stream. While generally lower in efficiency than rotary pumps, jet pumps need no moving solid parts. Related to jet pumps are other devices in which one flow induces another. A vacuum pump that attaches to a tap water outlet is a kind of inverse version—the main flow draws air out of the orifice. Old-fashioned carburetors drew in gasoline this way, and we sometimes ventilate buildings by using ambient wind to draw air through themselves.

Two points motivate this focus on the two categories. First, displacement pumps operate best at higher and dynamic pumps at lower pressures. Not that a sharp value of pressure marks the transition—the simplicity of displacement pumps makes them preferable for some low-pressure applications and the smooth operation of dynamic pumps underlies their use, often with multiple stages, to produce high pressures. And second, precisely these same distinctions of modus operandi and character of output apply to the pumps found in organisms. Whatever its imperfections, this mechanical and operational dichotomization can help us understand the particular distribution of pump types we observe in nature—an analytic and functional categorization complementary to our traditional phylogenetic viewpoint.

I'll begin by using the ways agrarian societies pump water to illustrate pump types in starkly basic forms, and with that basis tackle the greater complexity and diversity of living pumps. In engineering textbooks, fluid statics precedes fluid dynamics because of the relative simplicity of the former. Similarly, humans adopted displacement pumps before dynamic pumps, and displacement pumps remain more common in less industrialized technologies. Not that they lack diversity. One gets a good view of the different types from the pictures collected by Thorkild Schiøler and posted on the website of the Experimentarium, of Hellerup, Denmark (www.experimentarium.dk/uk/naturvidenskab_og_teknik/schiolers/); additional material can be downloaded from www.timsmills.info/URL-S/Animal%20Powered%20Systems.pdf.

Devices immediately recognizable as displacement pumps range from simple buckets and scoops to more complex pot chains, dragon-bone chains, swinging canoes (dhoons), shadufs, saquiyas (or saqias, etc.), and hoists (delous). One that, unusual for displacement pumps, avoids pulsatile or intermittent operation is the Archimedean screw (figure 10.2), a helical screw either turning in a tube or fixed within a cylindrical housing that turns. A Roman fresco at Pompeii shows an Archimedean screw pump driven as a treadmill by a person on the outside; in modern versions the operator usually turns a crank at one end. (At least in Europe cranks are medieval and later.) The shallow incline of the tube allows pockets of water to form; turning then raises the pockets. They still find occasional and even large-scale use at air-water interfaces—if fully submerged, pockets do not form, so they then become inefficient viscosity-dependent dynamic pumps.

The only fluid dynamic pump that seems at all common in preindustrial societies is the noria (figure 10.2), and it relies on both displacement and dynamics. A flowing stream turns an undershot waterwheel (the dynamic part); water-holders attached to the periphery of the same wheel fill from the same stream and raise water to an elevated spillway (the displacement part). As we'll see—and a reason to start with old human technology—nature seems to face similar (or at least analogous) difficulty in devising dynamic pumps. Indeed, judging from our technological versions, nature may face an additional obstacle. All but a few dynamic pumps (such as jet pumps) employ rotating elements, hence the common name "rotodynamic." And eucaryotic organisms don't make continuously rotating wheel-and-axle devices.

In looking at the pumps organisms employ, we'll limit our purview to those that move liquids, that is, water plus aqueous solutions and suspensions. As a less obvious circumscription, we'll for the most part exclude devices that move pumper rather than pumped, drawing an arbitrary line between, for instance, paddle-based locomotion and paddle pumps.

Living Displacement Pumps

The displacement pumps of organisms range from ones with close technological analogs to others that, although not fundamentally novel, have limited appeal for humans. Either we have more attractive alternatives or they work in ways awkward for our materials, machines, and applications.

Valve-and-chamber pumps. A single chamber whose volume can be changed, together with a pair of valves, satisfies the minimal requirements for such a pump. Our hearts, paradigmatic examples, have four chambers and six valves and operate as a pair of pumps, each with a two-stage pressure booster. Additional valve-and-chamber pumps return

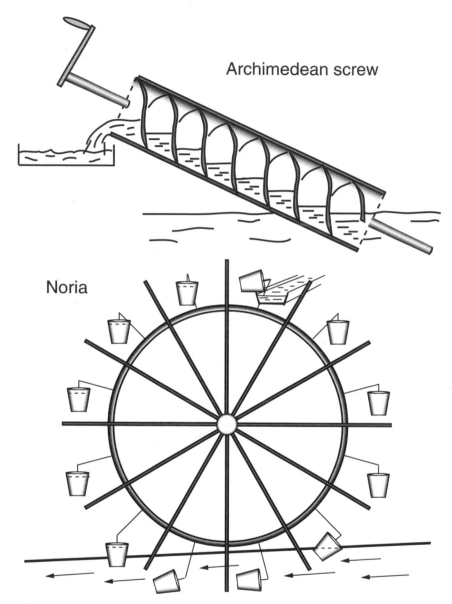

Archimedean screw

Noria

Figure 10.2. Two ancient devices for lifting water, an Archimedean screw and a noria; both are at least in part dynamic pumps. We still use a large-scale, modern version of the Archimedean screw for some low-lift, high-volume applications. But the noria has become anachronistic, most directly replaced by the hydraulic ram, which became practical in the nineteenth century with the advent of inexpensive metallic components of decent precision.

both blood from the veins of our legs and lymph from our tissues to our hearts—routine contractions of our skeletal muscles squeeze the chambers, these no more than the lengths of vessel between adjacent valves.

Valve-and-chamber hearts are widespread among the metazoa, particularly among vertebrates and mollusks. They require only ordinary additions—just valves and muscle—to ubiquitous flexible, tubular elements. Even single chambers can produce pressures of over 20,000 pascals, as do the left ventricles of tall mammals. Where skeletal or body-wall muscles can't be coopted, vessels may become not just valved but muscular and self-contractile, as are the lymph hearts of fishes, amphibians, and reptiles (Prosser 1973; Ottaviani and Tazzi 1977), and the lateral hearts of the giant earthworm, *Glossoscolex* (Johansen and Martin 1965). Some insect hearts may also operate as valve-and-chamber hearts (Jones 1977). Nor do all such pumps occur in circulatory systems. Valves on the inputs to the mantle chambers of jet-propelled cephalopods together with self-valving output funnels work as analogous chambers-with-paired-valves. Jetting scallops use their mantle cavities and mantles and as chambers-with-valves in a similar way, as do fishes that pump water over their gills using their mouths, oral and branchial chambers, and opercula (Lauder 1980).

Valveless chamber and piston pumps. Where pumps need produce only single pulses of fluid or reciprocating flows, valves become superfluous. That happens in many systems and appears in so many guises that the underlying commonality easily escapes notice. Our urethral pumps (see Glemain et al. 1990) work this way. Most jet propulsors, from pulsing jellyfish (DeMont and Gosline 1988) to the anal jets of dragonfly nymphs, are valveless chamber pumps. Anal jets can produce both single jet pulses (Hughes 1958) and repetitive respiratory flows (Pickard and Mill 1974), with the same equipment serving both functions. Similarly, most injectors make use of valveless chambers, including the venom injectors and squirters of rattlesnakes and cobras (Kardong and Lavin-Murcio 1993; Young et al. 2003, 2004), of the toxic snail, *Conus* (Schulz et al. 2004), and of spiders (Yigit et al. 2004). So does the branchial chamber of jetting fishes, another system that's respiratory as well (Brainerd et al. 1997).

And most suckers, including both blood- and nectar-sucking insects (Kingsolver and Daniel 1995), use valveless chambers. Even aphids, which can use the considerable hydrostatic pressures of plant phloem to drive fluids in through their stylets, retain the capacity to generate suction in this way (Kingsolver and Daniel 1995). Most often sucking chambers depend on expansion tied to the elastic recoil of some muscularly stressed material. Some sucking insects can produce pressures well below ambient, even, in the bug *Rhodnius*, truly subzero or negative pressure (Bennet-Clark 1963).

Less common than valveless chambers are piston pumps, perhaps because the physical arrangement represents something unusual in nature—however ordinary it might be in human technology. Several kinds of infaunal marine worms seem to irrigate their burrows by acting as piston pumps, in particular the clam worm *Nereis,* and the parchment worm *Chaetopterus.* In both cases, though, that may oversimplify the well-coordinated movements of appendages as well as body walls (Riisgård and Larsen 1995).

Valveless moving chamber (peristaltic) pumps. These typically produce pressures lower than the previous two types, and they're likely to be lower in energetic efficiency. But like them, they should be easy to evolve from a basic muscle-enclosed tube. Our intestines and our esophagi, of course, depend on peristalsis, as do the hearts of most annelids, holothurians (sea cucumbers), and arthropods (Martin 1974). Burrow irrigation in *Nereis,* noted above, seems to involve some peristaltic body wall movement as well as piston action. Inasmuch as earthworms locomote in an essentially peristaltic mode, we might expect similar peristaltic pumping among burrow-dwelling aquatic oligochaete as well as polychaete annelids.

Besides their undoubted ease of evolution, peristaltic pumps have functional advantages. Pumping liquids of high viscosity or with a lot of suspended solids presents no great problem. Peristaltic action provides mixing as well as lateral transport, offsetting the laminarity of low-speed flows in small pipes. That should matter where (as in intestines) absorption or exchange across pipe walls accompanies lengthwise transport—if not in, say, ureters. And pumping direction can be reversed with nothing more than a minor shift in neural or neuroendocrine signaling. Reversibility of our intestines provides a nonscatalogical closing for a somewhat heavy-handedly humorous book (Chappell 1930). Cud-chewing bovids routinely reverse their esophageal pumping, and insect hearts often switch directions (Jones 1977).

Osmotic pumps. These uniquely (but not necessarily) biological devices operate not by decreasing the size of a compartment but by increasing the volume of what the compartment must contain. Organisms rarely engage in active transport of water; instead, they move ions or small molecules, with water following by passive osmosis. Thus, the driving force for osmotic pumps comes from such indirect water transport or else from local increase in osmotic strength as a result of dimer, oligomer, or polymer hydrolysis.

Most osmotic pumps are small, and these are the predominant pumps driving bulk fluid flow in unicellular systems. However small the volumes they move, their pressures can be immense; that's fortuitously complementary with the peculiar ability of small systems to resist great pressure differences. A mere molar difference in solute concentration (assuming a non-electrolyte) across a membrane produces about 2.2 megapascals of pressure. To cite a specific case, the fungus *Gibberella* (Trail et al. 2005), about which more below, produces a peak pressure of about 1.5 megapascals. Nonetheless, these pumps suffer from several drawbacks. For one thing, they act at surfaces, so scaling up three-dimensionally takes a disproportionate increase in surface area, either with folds or villi or by proliferation of the basic units. For another, pumping ordinarily consumes the osmolyte, not as fuel, but by simple dilution—a water pump that works by hydrolyzing starch into osmotically active mono- or disaccharide will first dilute and then carry away the product in the flow it generates. So resynthesis may entail more than just metabolic reversal. While countercurrent or other such conservative devices may help, the basic problem cannot be entirely evaded.

Osmotic pumps figure in at least two of the schemes for throwing fungal projectiles that were described in the third chapter, those of *Pilobolus* and *Gibberella*. These osmotic engines power hydraulic ejection both by providing hyphae with liquid (volume) and by stretching their elastic walls (pressure); *Gibberella* does so by transporting potassium, with chloride coming along as counterion. Another osmotically charged hydraulic engine closes the Venus flytrap (Forterre et al. 2005). One can point as well to the excretory organs of animals, varying from partially osmotic to fully osmotic ones such as the aglomerular kidneys of some marine teleost fishes. (Our own kidneys capitalize on arterial blood pressure and thus on our hearts to drive their initial glomerular ultrafilter.)

The protonephridia of some acoelomate invertebrates, best known as flame cells in planaria, are a peculiar case—or possible case. Ducts opening to the exterior remove excess water, as required by these leaky freshwater creatures. Presumably water transport follows some osmolyte secretion that gets reabsorbed. One wonders about the role of the one cilium (in solenocytes) or the tuft of cilia (in flame cells) at the blind ends of these ducts (Schmidt-Nielsen 1997). Osmotic pumps work at high pressures and low flow rates, while ciliary pumps, as fluid dynamic devices, do best doing opposite service. I've seen no suggestion about what good—perhaps a bit of stirring—one or a few cilia might do under such circumstances. Our own renal tubules may have cilia, but no analogy should be drawn. Ours lack central microtubules and can't propel fluid; instead they appear to work conversely, as signal generators, in particular as flow sensors (Yokoyama 2004).

Osmotic pumps play major parts in two large-scale fluid transport systems, although in both instances the mechanistic details aren't fully elucidated. If you cut off the top of a well-watered herbaceous plant, sap oozes out from the xylem elements on the cut surface. Water, absorbed from the soil, has been pumped up the stem by so-called root pressure. Herbaceous stems may provide its most obvious expression, but it occurs in some large, woody plants as well. Pickard (2003a, 2003b) provides a good view of the present knowns and unknowns concerning root pressure.

While flow in xylem depends mainly on evaporative pull from the top, osmotic pumping seems mainly responsible for flow in the complementary tissue, phloem. Again, the details have given trouble. The classic Munch hypothesis from the 1930s invokes osmotic forces, and they clearly play some role. But here too we lack the full picture; here we encounter a particularly daunting diversity of structures, flow pathways, and chemistry. A look at, for instance, van Bel (1993) or Thompson and Holbrook (2003) will give some sense of the problems.

A good osmotic pump should get as much passive water movement as possible for a given amount of osmolyte transport. That explains a common feature among such pumps. Instead of secreting osmolyte into large external (or extracellular) spaces, they discharge it into restricted areas, diffusively remote from those larger volumes. Thus its concentration will be (and for a time will remain) higher. Depending on the system, osmolyte may be ultimately lost downstream or actively reabsorbed for reuse. The loops of Henle of mammalian kidneys play a particularly fancy version of this game, with a countercurrent multiplier isolating a region of high osmolarity. Recognition of the basic arrangement traces to Curran (1960), in rat intestines, and to Curran and MacIntosh (1962) as a general phenomenon. It was later demonstrated in the water-ejecting invaginations of gall bladder cells by Diamond and Bossert (1967), who gave it the nicely descriptive name "standing gradient osmotic flow." As noted by Tyree and Zimmermann (2002) (and at least tacitly by Nobel 2005), such standing gradient osmotic devices complement evaporative pumps in higher plants. It plays crucial roles in generating root pressure and in loading and propelling flow in phloem—phenomena just mentioned.

Evaporative pumps. In one sense an evaporative pump is the opposite of an osmotic pump. Instead of generating positive pressure by transport of osmolytes and water into a compartment, it produces negative pressure by removing liquid from a compartment. Like osmotic pumping, evaporative pumping requires no macroscopic moving machinery, preadapting it for use by plants. Osmotic pumping depends on differentially permeable membranes, biologically ubiquitous; evaporative pumping requires an air-water interface, limiting its applications to terrestrial or semiterrestrial organisms. So evaporative pumps should be less widely distributed. They should also be limited by a peculiar asymmetry between positive and negative pressures. While pressures can be increased without intrinsic limit, they can't easily be decreased much below zero. Thus one expects that a pressure drop could only offset ambient pressure. A further constraint is that evaporation must occur across a surface that can tolerate the pressure difference that it generates.

Despite such limitations, evaporative pumping probably moves more liquid through organisms than do all other macroscopic pumps combined. One might say it does the heavy lifting, drawing water from soil and raising it to the photosynthetic structures of terrestrial plants. In some sense such water loss is the price these pay to obtain sparse and precious atmospheric CO_2. Plants evade the zero pressure limit, not trivially but monumentally, generating tensions in water as low as −12 megapascals (−120 atmospheres), far, far below zero. They contain interfaces that withstand such pressure differences without either collapsing or restricting evaporation—combining the fine-scale cellulose meshwork of their cell walls and the high surface tension of water in contact with air. The seventh chapter described this remarkable scheme. Here just note that the evaporative pumps of terrestrial plants produce the most extreme pressure differences of any biological pumps, and that such pumps are rare elsewhere in either natural or human technology.

NATURE'S DYNAMIC PUMPS

For these we might better use the additionally qualified name, *fluid dynamic pumps*, because another group of biological pumps depends on the dynamics of solid materials. Compared to the analogous devices in human technology, the fluid dynamic pumps of organisms appear less diverse. They're also more distant from our devices as a consequence of two basic differences between the two technologies—nature's inability to make macroscopic rotational machinery and our lack of anything much like cilia or ciliated and thus wall-pumping tubes.

Drag-based paddles. In our quest for efficient propulsion, propellers, which move blades normal to flow, have almost entirely replaced paddles, which move them parallel to flow. One must go back to the noria, a preindustrial irrigation device, to find a fluid dynamic pump based on the drag of broad blades in flow. Similarly, nature makes only limited use of pumps based on paddling. Foster-Smith (1978) recognized such a pump in the amphipod crustacean, *Corophium volutator*, which burrows in mud and propels water with reciprocating pleopods. I suspect that members of the infaunal shrimp genera *Upogebia* and *Callianassa* do likewise. But they probably do so only occasionally, depending in part on flow induced by asymmetry of the apertures of their U-shaped burrows.

Foster-Smith found that *Corophium* could generate pressures only about 4 percent as high as those made by the piston pumpers *Nereis* and *Arenicola*, although for its size it could drive considerably greater volume flows.

Lift-based propellers. My search for liquid-propelling pumps, *sensu strictu*, that use propellers has come up nearly dry. Some fishes do ventilate egg masses by beating their tails while stationary, but I can't cite specific performance data. In air one can point to a hive-ventilation system used by honeybees. One or a series of honeybees beat their wings while standing just beyond the entrance to their hive. Hertel (1966) points out that a line of bees constitutes a multistage axial compressor analogous to that used in the jet engines of aircraft. (But the photograph he provides has been inappropriately retouched.) Southwick and Moritz (1987) claim that the hives "breathe" as the bees alternately pump it out and allow it to inhale elastically. The present discussion of pumps suggests otherwise—even a line of bees should form a high-volume, low-pressure pump, and beehives do not feel as if they have a low enough elastic modulus and high enough resiliency.

Ciliated surfaces and chambers. By contrast with both these fluid dynamic pumps, ciliary pumps abound. Muscle must be persuaded to move fluid with some form of transducing equipment; cilia do so as their basic modus operandi. Cilia may be far slower in operation than muscle, but collecting manifolds with decreasing aggregate cross-sectional areas can raise the output velocity of ciliary or flagellar pumps. For instance, sponges eject water with their flagellar pumps at about 0.2 meters per second, a far higher speed than any flagellum can generate directly. Ciliary pumps find wide use in low-pressure, high-volume applications such as suspension feeding.

But they have several drawbacks. Cilia are microscopic and work at that scale. So scaling up a pipe with ciliated walls bumps against a pumped cross-section that increases faster than the pumping circumference. In addition, the cilia-lined pipe can't have the gently parabolic velocity gradient of a remotely pumped pipe. The entire gradient from the mandatory zero speed at the wall to the peak speed of the pipe can't span as much as the length of a cilium, so the gradient becomes severe even at modest flow speeds. Since viscous power loss depends on the steepness of that velocity gradient, ciliary pumping suffers from an intrinsically low efficiency in all but the narrowest pipes and channels. At least that steep velocity gradient will help when organisms exchange material or heat across pipe or channel walls. Thus, ciliated surfaces serve admirably for organs such as gills—as in gastropod mollusks, where they pump water for respiration, and in bivalve mollusks, where they play a central role in suspension feeding (as noted in chapter 1).

One wonders whether this inauspicious scaling explains their absence on the gills of fish. One also wonders whether they'd be useful on the gills of aquatic arthropods were motile cilia known to that phylum. Less puzzling is their absence as pumps in our capillaries. Velocity presents no problem, since blood speeds are appropriate for ciliary pumping. But effective operation is precluded by the relatively high resistances of circulatory systems. In the view given in figure 10.1, the steep slope of a system resistance line from the origin of that application mates poorly with the low pressure maximum of any ciliary pump capability envelope. We might wish for circulatory systems in which ciliated capillaries make our fallible hearts unnecessary. But that low pressure-generating capability rules them out, at least where blood volumes remain under 10 percent of body volumes, as in both vertebrates and cephalopods. (La-Barbera and Vogel 1982 failed to consider the need to match pump performance with system resistance, so we attributed the choice between ciliary and muscular pumping solely to ancestry—but we've learned better.) For the same reason, as noted earlier, one suspects that pumping can't be the primary function of the cilia of flame cells and solenocytes.

Capillary (surface-tension) pumps. Inasmuch as it lifts water against gravity, the capillary rise of water in a narrow hydrophilic tube constitutes a proper pump. As does evaporative pumping, such surface-tension pumping works only with an air-water interface; so, similarly, it's available only to terrestrial and semi-aquatic organisms. We may know a considerably wider range of cases than for evaporative pumping, but by comparison its role remains a modest one.

From time to time one runs across statements (by nonbiologists) asserting that sap rises in trees as a result of capillarity—simply by the ascent of an aqueous fluid in a hydrophilic tube. That can't be the case, as repeatedly pointed out (see, for instance, Nobel 2005), because the conduits are just too wide. The capillary rise of water in a circular vertical tube, h, is

$$h = \frac{2\gamma \cos\theta}{\rho g r}, \tag{10.1}$$

where γ is its surface tension, 0.073 newtons per meter; ρ its density, 1000 kilograms per cubic meter; θ the contact angle (0° for perfect wetting); g is gravitational acceleration; and r is tube radius. Even under ideal circumstances, water will rise only 1.5 meters in a small tracheid, one 20 micrometers in diameter. In a xylem vessel of 200 micrometers, as in an oak, the rise will be ten times less. Even with perfect wetting, a 50-meter column of water could be sustained by capillarity in no tube over 0.6 micrometers across.

Capillarity does matter here and there, for the most part with narrow tubes of no great length. Rehydration of dry stems and leaves of the resurrection plant *Myrothamnus* depends on it, but conduit diameters are of the order of 2 micrometers (Schneider et al. 2000; Tyree 2001). Some insects, most notably orchid bees, draw in nectar through their probosci at least in part by capillarity (Kingsolver and Daniel 1995; Borrell 2004). At least two kinds of birds use the mechanism, hummingbirds to draw in nectar (Kingsolver and Daniel 1983) and phalaropes to raise small quantities of water with edible plankton up a vertical bill whose tip has been dipped in a body of water (Rubega and Obst 1993).

In the guise of wicking, capillarity can move liquid upward on the outsides of sufficiently wettable surfaces. A few cases have been described, not surprisingly, in amphibians. Lungless plethodontid salamanders of the genus *Desmognathus*, which breathe through wet skin, can stay wet by wicking water upward as well as by exuding body water through that skin (Lillywhite 2006). Toads (genus *Bufo*), which lack skin mucus, can stay moist by wicking as well (Lillywhite and Licht 1974).

Flow inducers. These may be more common in nature than in modern human technology. While our jet pumps have little in the way of immediate natural analogs, nature capitalizes both on the elevated pressure of oncoming flows and on the reduced pressure due to flow over an orifice that opens normal to that flow. Their low pressures impose the main limitation on flow-inducing pumps. Pressure cannot deviate either upward or downward from ambient by significantly more than the dynamic pressure difference, Δp, defined by Bernoulli's principle,

$$\Delta p = \frac{\rho v^2}{2},$$ (10.2)

where v is the speed of flow. (For small-scale and thus low Reynolds number flows, pressures will be still lower.) A fish swimming steadily at 0.5 meters per second can generate only about 125 pascals, or an eight-hundredth of an atmosphere. A suspension feeder working at 0.1 meters per second has available a mere 5 pascals, or a twenty-thousandth of an atmosphere. Still, a baleen whale swimming with open mouth at 3 meters per second can take advantage of a more substantial 4500 pascals, or a twenty-secondth of an atmosphere. I gave considerable space to flow-inducing schemes earlier (Vogel 1994b) and will just mention representative cases here.

Ram ventilation in fishes is the best-known case of pressure elevation at an orifice facing into a flow. Its use varies fish to fish, from a trivial role for it in some, to a role only at high swimming speeds (where respiratory needs are greatest, of course) in others, to total dependence on it in some large, fast fish that must either swim or asphyxiate (Steffensen 1985). Analogous pumps with upstream-facing inputs drive the suspension feeding systems of some ascidians and caddisfly larvae. They probably drive fluid through the olfactory passages of many fish as well (Cox 2008).

Still lower pressures are available for drawing fluid out of an elevated orifice. The arrangement, though, finds use by keyhole limpets and abalones to draw water across their gills for respiration. With it some sand dollars draw food-laden water from underlying sediments up past their oral surfaces and out through their oral-aboral slots. With it the shrimp genera mentioned earlier probably irrigate their U-shaped burrows. Sponges take advantage of both elevated pressures on their upstream-facing (and indirectly on their other) ostial inputs and the reduced pressures at the oscular outputs.

Temperature gradient pumps. Flows can be induced in several ways by spatial variations in temperature as noted in chapter 4. Most such pumps move air rather than water. Free convection, the most obvious, drives the internal circulation of some giant African termite mounds (Turner 2000) as well as providing some cooling currents around sunlit trees during periods of unusually low wind. An arrangement related to evaporative sap-lifters pumps by evaporation in one place and condensation in another—as in heat pipes—and constitutes a kind of dynamic pump. But at this point its use by organisms is no more than a suspicion. Finally (and more clearly dynamic) is Marangoni pumping, flow driven by surface tension gradients that follow temperature gradients—again, it remains something to be kept in mind as a distinct possibility for organisms. All of these pumps develop only the lowest of pressures.

An Index for Pump Performance

Figure 10.3 gathers data for pressure boost and volume flow for fifty-three pumps—thirty-seven displacement and sixteen dynamic. They were chosen for their diversity in function and the range of values of the two variables; they include sap-lifters, hearts, blood suckers, jets, projectile ejectors, gill irrigators, and suspension feeders. The graph seems to confirm the generalization that in nature as in human technology, displacement pumps for the most part work at higher pressures and dynamic pumps at lower pressures; a *t*-test of the data gives a significance level for that distinction of about $p=0.05$. The data support considerably less well the notion that displacement pumps work at lower volume flows, either by inspection of the graph or by another *t*-test.

Might a single indicator encapsulate our picture of how biological pumps sort out? One might calculate a ratio of the two variables, $\Delta p/Q$,

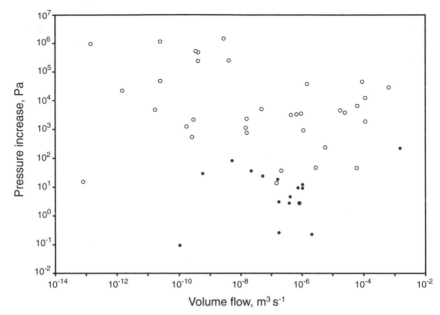

Figure 10.3. Pressure produced versus resulting volume flow for a collection of pumps; all of those of figure 10.4 are included, plus others (of the same general types) for which I lacked data for either radius or flow speed. Open circles indicate positive displacement pumps, solid circles dynamic pumps.

for each of the pumps on the graph—or for any others. The mean $\Delta p/Q$ of the displacement pumps does come out higher than that of the dynamic pumps. But, like the data for volume flow, the distinction between our two general pump types fails statistical scrutiny by a clear margin ($p > 0.2$). So the ratio has little if any value for biological pumps. Most likely, the trouble comes from the huge size range of these pumps together with the intrinsic size-and speed-sensitivity of the ratio. Thus, dimensional manipulation of the ratio gives $\rho v/S$, where S is conduit cross-section. Pumping a given medium, faster flowing systems will be biased toward higher values, while larger systems tend toward lower values. A dimensionless ratio might be more informative, assuming one can be found with unambiguously defined and easily determined variables.

One possible ratio is the pressure coefficient, C_p, long used in fluid mechanics to describe pressure distributions around bodies in flow. It divides pressure by dynamic pressure, the pressure that would be generated were the moving fluid to be suddenly halted, as described by Bernoulli's principle. Specifically,

$$C_p = \frac{2\Delta p}{\rho v^2},$$ (10.3)

in effect a dimensionless form of equation (10.2). v represents the highest speed in the system, most often at the output of the pump. Pressure change appears, as we would like, in the numerator, while flow speed stands in for volume flow in the denominator.

In practice, unfortunately, the pressure coefficient does almost as badly at sorting out pump types as a pressure-volume flow ratio. Applying it to thirty-nine of the previous fifty-three pumps for which I found adequate data yields a distinctly odd ranking. For instance, both the lymph hearts of toads and our own lymphatic vessels, which we mainly power with our skeletal muscles, have values up with the xylem of trees, a bias attributable to their very low speeds. And the sporangium of the fungus, *Pilobolus*, an osmotic engine, gives the lowest value of all; its very high speed overcompensates for its substantial pressure. At least the ratio puts the very high-pressure, low-flow xylem of pine, oak, and the vine *Entadopsis* near the top. Its oddities probably come from the tacit presumption in the formula of flow at high Reynolds numbers—relatively large, fast, and turbulent—rather than at biologically appropriate low Reynolds numbers with their laminar flows. (One might add a subscripted t for "turbulent" and designate this pressure coefficient C_{pt}.) Put another way, its denominator reflects inertial energy loss rather than a more relevant viscous loss.

In my earlier look at pumps (Vogel 1995a) the coefficient was for just this reason replaced by one that presumed viscous rather than inertial pressure loss,

$$I = \frac{\Delta p t}{\mu},$$ (10.4)

where t is time and μ is viscosity. Density has deferred to viscosity, as usually happens in low-Reynolds-number formulas. Among a set of pumps more limited than the present one, xylem and hearts came out at the top, as we think they should. But two displacement pumps produced the lowest indices, the jet of the jellyfish *Polyorchis* and the blood sucker of the bug *Rhodnius*. This last generates the greatest known pressure difference in an animal. The index has a practical problem as well, the interpretation of t, a kind of length-less inverse velocity. Without great conviction, I took it as the transit time for a bit of fluid to pass the part of the system with the greatest resistance. Not only does it take more guesswork than one would prefer but it cannot escape ambiguity when applied to the tapering pipes so common among organisms.

A dozen years later, I offer an alternative dimensionless ratio. This one divides the pressure force, pressure times cross-section, by viscous resistive force. The latter is the product of viscosity, flow speed, and vessel radius. It comes either from Stokes' law for the drag of a sphere or from an equation (equation 13.17 in Vogel 1994b) for the pressure drop of flow through a circular orifice. We might call it the "pressure coefficient for laminar flow" to draw an analogy with the well-established (turbulent) pressure coefficient of equation (10.3). Specifically,

$$C_{pl} = \frac{\Delta pr}{\mu v}. \tag{10.5}$$

To get the ratio from what looks like the most appropriate source, the Hagen-Poiseuille equation for laminar flow through a circular pipe, one need only assume an isometry in which pipe length can be replaced by pipe radius. As we'll see, that assumption occasionally generates peculiar values, something that should be kept in mind when drawing inferences from values of the ratio.

Figure 10.4 gives values for the thirty-nine pumps previously mentioned— thirty displacement and nine dynamic. What can we make of these numbers? While the overall range of values varies 2.5-million-fold, functionally homogeneous groups cluster satisfyingly. The xylem pumps, the jets, and the blood suckers each span ten-fold ranges, while the hearts (including both single and dual-stage pumps) vary less than fifty-fold. Ciliary and flagellar pumps vary about 6.6-fold, with the flagellar one of a sponge not unexpectedly the lowest. Thus, the ratio might provide expectations for pumps not yet analyzed, in short, to have predictive value.

The evaporative pumps of xylem come out at the top, with the highest value for the narrower tracheids of gymnosperms—they generate comparable pressures but get less flow from them than do the wider vessels of broad-leafed trees. A vine, with (as is typical) the widest vessels, has the lowest value. r may be larger than for the others, but v increases by a greater factor.

Flows over the gills of fishes span a wider range, 360-fold, but the distribution of cases within that range looks quite reasonable. Suction feeding, with its necessity for rapid, impulsive flow, yields the highest value. The ram-ventilating tuna also has a high ratio, the highest for any dynamic pump. But it swims exceptionally fast, and so has access to the greatest driving pressure. One also sees a low value for actively ventilated respiratory gills, consistent with a function demanding continuous, low-cost flow; one suspects that the low value preadapts the system to take advantage of ram ventilation during faster swimming.

Two dynamic pumps, both of suspension feeders, give values that look anomalously high, the paddle pump of the amphipod *Corophium* and

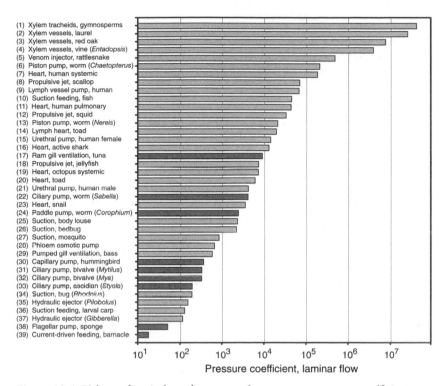

(1) Xylem tracheids, gymnosperms
(2) Xylem vessels, laurel
(3) Xylem vessels, red oak
(4) Xylem vessels, vine (Entadopsis)
(5) Venom injector, rattlesnake
(6) Piston pump, worm (Chaetopterus)
(7) Heart, human systemic
(8) Propulsive jet, scallop
(9) Lymph vessel pump, human
(10) Suction feeding, fish
(11) Heart, human pulmonary
(12) Propulsive jet, squid
(13) Piston pump, worm (Nereis)
(14) Lymph heart, toad
(15) Urethral pump, human female
(16) Heart, active shark
(17) Ram gill ventilation, tuna
(18) Propulsive jet, jellyfish
(19) Heart, octopus systemic
(20) Heart, toad
(21) Urethral pump, human male
(22) Ciliary pump, worm (Sabella)
(23) Heart, snail
(24) Paddle pump, worm (Corophium)
(25) Suction, body louse
(26) Suction, bedbug
(27) Suction, mosquito
(28) Phloem osmotic pump
(29) Pumped gill ventilation, bass
(30) Capillary pump, hummingbird
(31) Ciliary pump, bivalve (Mytilus)
(32) Ciliary pump, bivalve (Mya)
(33) Ciliary pump, ascidian (Styola)
(34) Suction, bug (Rhodnius)
(35) Hydraulic ejector (Pilobolus)
(36) Suction feeding, larval carp
(37) Hydraulic ejector (Gibberella)
(38) Flagellar pump, sponge
(39) Current-driven feeding, barnacle

10^1 10^2 10^3 10^4 10^5 10^6 10^7
Pressure coefficient, laminar flow

Figure 10.4. Values of an index of pump performance, a pressure coefficient that applies to laminar flows. Darker bars mark the dynamic pumps. [Sources: (1) Cermák et al. 1992, Pittermann and Sperry 2003; (2) Cermák et al. 1992; (3) Kramer 1959, Zimmermann 1971; (4) Fichtner and Schulze 1990; (5) Kardong and Lavin-Murcio 1993, Young et al. 2003; (6, 13, 22, 31, 32, 33) Riisgård and Larsen 1995; (7, 11) textbook values; (8) Cheng et al. 1996; (9) Milnor 1990; (10) Alexander 1969; (12) Shadwick 1994; (14) Müller 1833, Jones et al. 1992; (15, 21) Glemain et al. 1990; (16) Lai et al. 1990; (17) Stevens and Lightfoot 1986; (18) DeMont and Gosline 1988; (19) Wells 1987, Agnisola 1990; (20) Gibbons and Shadwick 1991; (23) Jones 1983; (24) Foster-Smith 1978; (25, 26, 27) Daniel and Kingsolver 1983; (28) Nobel 2005; (29) Lauder 1984; (30) Kingsolver and Daniel 1983; (34) Bennet-Clark 1963; (35) chapter 3; (36) Drost et al. 1988; (37) Trail et al. 2005; (38) Bidder 1923, Vogel 1978; (39) Trager et al. 1990.]

the ciliary pump of the polychaete *Sabella*. But we may be miscategorizing the infaunal *Corophium* as a dynamic pump. Since the paddles operate within its tube they work more like a set of moving compartments, as in the human- or wind-powered "dragon-bone" pumps in East Asia that move water from one rice paddy to another. They're less closely analogous to the serial paddles on say, a rowed trireme or galley. *Sabella*'s ratio may draw attention to the limitation of the ratio alluded to earlier. It erects a fan of ciliated tentacles normal to flow, in effect a huge number of ciliary pumps operating in parallel. That parallel array makes use of the radius of the entire array, as done here, at least questionable. Using the distance between individual ciliated elements reduces the value down to the level of the other ciliary pumps, those of two bivalves and an ascidian. I've not altered either the categorization of *Corophium* or the value for *Sabella*, in part to preserve them as illustrative examples and in part to avoid conscience-troubling *post hoc* adjustments.

A look at refilling in a sea anemone points up the predictive value of the ratio especially well. When disturbed, *Metridium* can contract its body wall musculature and collapse into an inconspicuous flat blob, largely by expelling almost all the water in its central gastrovascular cavity through its mouth (Batham and Pantin 1950). According to textbook accounts, it slowly reinflates by pumping water back in through ciliated tubes, the siphonoglyphs, while it keeps its mouth closed. Batham and Pantin (1950) measured reinflation pressures of around 25 pascals. Using the dimensions of their animals and an estimate (from various sources and personal observations) of an hour for the process, I calculate a pressure coefficient for laminar flow of 1.0×10^5. The value lies between those of our systemic hearts and of the jets of scallops, about 250 times higher than even the exceptionally high value for the ciliary pump of *Sabella* and a million times higher than typical ones. Perhaps the usual accounts incorrectly assume that pumping by and through the ciliated siphoglyphs does most of the work. So prompted, one looks and finds easily another player. The viscoelasticity of the body wall of *Metridium* has about the right elastic modulus and temporal behavior for the task, judging from the measurements of Alexander (1962), Gosline (1971), and Koehl (1977). In short, elastic recoil may drive refilling.

Several final notes on this measure of pump performance. Notice that the ratio contains the product of pressure and radius. Since for a given material the tolerable pressure varies inversely with the radius of the pipe, this nicely offsets any size-related bias that Laplace's law might impose. The ratio does less well in correcting for the effects of collecting or expanding manifolds—constricting a flow will increase its speed, both lowering the numerator and raising the denominator.

Perhaps unexpectedly, whether or not pumping is sustained makes little difference. Venom injection by rattlesnakes and spore ejection by fungi are quick, one-shot mechanical processes, but their widely different values span almost the entire range of long-acting pumps. Some blood suckers (like *Rhodnius* and lice) remain painlessly attached to their hosts for long periods, others (like mosquitoes) get the job done quickly lest they be swatted; nonetheless, their ratios differ little. Such indifferences suggest that efficiency, energy, and power play secondary roles in determining the match of pump type to application. That parallels what I noted for the scaling of ballistics in the second chapter, where force (there in the guise of stress) appeared more critical than work or efficiency.

YET ANOTHER KIND OF PUMP

Earlier I noted in passing the imperfection of the dichotomy between displacement and dynamic pumps in human technology. That same imperfection surfaced for nature's pumps in the burrow-ventilation of an amphipod that used paddles, ordinarily dynamic, to form moving compartments, as in some displacement pumps. Distinguishing the two categories among the pumps of organisms may bump into a more general difficulty, the exclusion of yet other categories of devices. Perhaps we should bear in mind that biomechanics usually recognizes functional devices in nature when they have analogs in human technology. Clearly not everything in our technology has a natural equivalent. Less obviously, devices in nature may lack technological analogs—less obviously because that lack makes them easier to overlook.

In their structural materials, the two technologies differ in their relative reliance on rigid and flexible ones. When we speak of artifacts that change shape under load as "deformed" we adopt a word whose pejorative sense reflects our human preference for rigidity. Think back—only two of the technological pumps mentioned make significant use of flexible materials. Peristaltic pumps squeeze tubing; while inefficient, they avoid possible contamination of fluid with pump parts. Diaphragm pumps pulse a (periodically replaced) rubber or neoprene membrane to change the volume of a compartment; they have some advantage when, as in sewage systems, pumping slurries of suspended solids. But in nature flexibility appears to be the default condition, with rigid materials appearing only where functionally mandatory. Beside peristaltic pumps, can we recognize others that depend on flexible materials?

When visiting the laboratory of Mory Gharib a few years ago, I was shown a more refined version of the device pictured in figure 10.5, one in which a very flexible element completes a circuit of much less flexible conduits. Repeatedly compressing the flexible tubing near one of its ends produces an impressively strong unidirectional flow without either check-valves or peristalsis. Depending on where the flexible tubing is compressed, this valveless pump can be run in either direction. How it works is clear enough—compressing the tube forces fluid in both directions, but one responds mainly by expansion of the flexible tubing rather than by

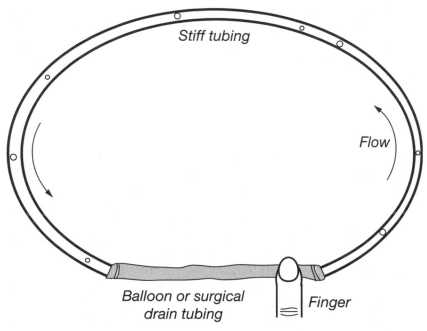

Figure 10.5. Perhaps the simplest possible solid dynamic liquid pump. Pressing repeatedly near the right end of the flexible tube, as here, drives the fluid counterclockwise around the circuit—as one can see with either a few trapped bubbles or some suspended matter. The pump tolerates addition of a substantial resistive element in the stiff tubing, and even a very crude one can generate 2500 pascals.

sending the fluid on through the rest of the circuit. Crude models tolerate a wide range of sizes, tube flexibilities, and circuit resistances.

The hearts of ascidians may work like these models. Their hearts have long been known to reverse periodically, accomplished by changing the end at which a pacemaker triggers constriction (Martin 1974). Rather than relying on reversible peristalsis, they depend on some other direction-determining arrangement. By cardiac standards, ascidian hearts produce only modest pressures, about 300 pascals in a large one (Goddard 1972), reflecting the modest resistance levels of their circulations. While insect hearts, also valveless low-pressure devices, commonly reverse, they have usually been described as peristaltic (Sláma 2003). Still, given the great diversity of the arthropods, this kind of valveless, nonperistaltic, reversing pump might well occur somewhere among them.

Forouhar et al. (2006) recently described something analogous to this valveless pump, one also unlikely to be limited to a small group of organisms. As they point out, an embryonic vertebrate heart begins pumping well before it develops valves. Working with zebrafish embryos, they showed that a tubular heart need not depend on peristalsis. Instead, they propose that pumping results from the suction generated by propagation of an elastic wave in the wall of the heart.

Perhaps we should entertain the idea that organisms make use of a third general category of pumps, one that might be called "solid-dynamic pumps," together with explicit use of the qualifier "fluid" for the dynamic pumps described earlier. Such solid-dynamic pumps would likely be associated with quite specific tunings of the multidimensional properties of flexible biological materials. We vertebrates and cephalopods certainly use something close to such a pump as we buffer the radical pressure fluctuations of our hearts with flexible arterial walls, walls with stress-strain curves tuned to our various blood pressures.

PERSPECTIVES AND SPECULATIONS

Two final items. First, we have here curious and varied combinations of functional and phylogenetic constraints. Higher plants must do with pumps that need no moving, macroscopic solid parts, which largely limits them to evaporative and osmotic pumps. Both such displacement pumps can generate wonderfully high pressures but neither shines when volume flow is the measure. Induction by external flows remains an option, but in terrestrial systems it will move air rather than water. Similarly, a few kinds of animals such as sponges can make no macroscopic pumping machinery, but at least cilia and flagella allow reasonable low-pressure, high-flow pumping. Arthropods know nothing of motile cilia, relying mainly on the movements of rake-like appendages for suspension feeding and on peristaltic and hydraulic pumping for internal bulk transport. Yes, the pumps of nature appear well-chosen for their assigned tasks, but no, no creature has anything approaching a free choice from a comprehensive catalog.

Second, whether analyzing locomotion, photosynthesis, or foraging, we biologists have given considerable attention to energetic or power efficiency as an index of performance. This chapter, for instance, stated early on that a maximal product of pressure generated and volume moved per unit time, power output, marked a pump as well matched to its task. Most often energetic efficiency can be unambiguously defined, and it sits well with prejudices from our physics courses, the physical devices in our lives, and our fuel bills.

But I'm skeptical about efficiency as a unique or even a particularly good comparative measure of devices of such disparate function as the pumps that move aqueous liquids through organisms. For one thing, all too often quantity of water moved may not adequately represent useful output. A suspension feeder may prefer to move less water if by doing so it can increase the fraction of edible material it extracts. Similarly, the cost relative to oxygen extracted by a gill may be minimized at a different flow rate than one that minimizes cost relative to the volume of water pumped.

For another, pressure ordinarily represents what we might call an unavoidable evil. A system may find itself stuck with a certain minimum from first principles, as with the gravitational loss of sap ascent. Or it may reflect some trade-off, as with circulatory vessel size versus the effectiveness of transmural diffusive exchange or with blood volume versus speed of flow. Only in hydraulic systems such as various fungal projectile ejectors does pressure matter as much as flow.

For that matter, metabolic cost must be the least immediate of desiderata for many pumps—what fraction of its overall output does a rattlesnake devote to squirting venom? Should a mosquito suck more slowly to minimize the cost of getting its dinner? We might assert (admitting the rare exception) that energetic efficiency will be a factor, if at all, for pumps that operate steadily rather than for those that need give only an occasional pulse.

Nor need pumping incur any metabolic cost at all. Only initial construction and maintenance impose any cost on, somewhat paradoxically, both the highest and the lowest pressure pumps—solar energy powers the evaporative pump that lifts sap while the energy of fluid moving with respect to a surface powers both ram ventilation and the variously employed current-induced flows.

In short, a look at pumps may inject valuable doubt about whether we can find in the living world a straightforward measure of utility comparable to power or energy in mechanical technology, information in telecommunications and computing, or money in economics.

CHAPTER 11

To Twist or Bend When Stressed

Probing the Proper Property

Several themes weave through the chapter that follows. While any—or all—might be left for a final encapsulation, perhaps they're better borne in mind while reading it.

- For all its arcane and counterintuitive phenomena, fluid mechanics builds its bioportentous aspects on just a few material properties of gases and liquids—density, viscosity, and sometimes surface tension. And we're mainly interested in just two substances, air and water. Solid mechanics, however great its intuitive familiarity, encompasses a daunting host of potentially significant material properties—three elastic moduli, three strengths, three maximum deformations, and three strain energy storages, corresponding to tensile, compressive, and shearing loads; up to six Poisson's ratios; work of fracture; hardness; density; and yet others. In addition, and of similar relevance, it cares about structural properties such as flexural and torsional stiffness. Still, for any given application, biological or techno-logical, only a few properties will directly affect functional success. Often an investigation must begin with a decision—or guess—as to which proper-ties might matter.

- As repeatedly pointed out by an engineer, the late James E. Gordon, humans typically design structures to offer adequate stiffness (plus, of course, a safety factor). He noted that nature, by contrast, appears to design for adequate (again plus some margin) strength, a criterion that ordinarily demands less material. The present discussion goes beyond that material economy to focus on the way a fundamentally different philosophy of design might depend on different suites of properties. We may not need to devise novel ways to describe materials, but we might well need to make different selections among the properties defined by mechanical engineers.

- While structural properties depend on material properties, they also de-pend on geometric properties—on shape. Put another way, they depend both on what goes into a structure and on its arrangement. Relying on pure materials, simple composites, or preexisting natural materials, we humans typically alter structural properties by tinkering with geometry.

Nature, adept at making materials whose properties vary from place to place within structures, appears to play as much with material as with geometry. Central to her structural technology are anisotropic composite materials—ones whose properties depend on load direction and location within structures. Perhaps the difference goes back to the contrast between Nature's factories, cells and their components, much smaller than the structures they produce, and our factories, which produce products smaller than themselves. But whatever the underlying cause, the availability of location-tuned, complex composite materials must affect structural designs.

Extending the argument takes a brief introduction to materials and structures. Materials can be stressed in three distinct ways—they can be stretched (tension), squeezed (compression), or sheared. Structures, in addition, can be bent, that is, loaded flexurally, or twisted, meaning loaded torsionally. Each of these structural loadings combines several material stresses. Thus, a beam extending outward from a vertical support responds to bending its free end downward with tension along and below its top surface, compression along and above its bottom surface, and shear in between, as in figure 11.1a. A shaft extending outward responds to twisting its free end with tension and shear on and near its outer surface and compression in the middle, as in figure 11.2a—we use that compression when wringing out a wet cloth with a twist. Textbooks first introduce tension, in practice the simplest stress, then compression, and finally shear. When considering structures, they look first at flexion and then at torsion. As a result, shear and torsion often get short shrift; as a further consequence, we too easily ignore what appeared as afterthoughts.

In part responding to this underemphasis, I want to focus on torsion. Never mind equal time or physical balance—I worry that we may overlook or misinterpret situations in which torsional stresses and their resulting strains play adaptationally significant roles.

In looking at such cases, we meet familiar problems of standards and scale. Compilations of shear moduli or torsional stiffnesses provide poor bases for evaluations, since they provide no obvious frames of reference for the biologist. Moduli refer to materials with no consideration of shape, and structural stiffnesses depend intrinsically on size. But both problems can be solved by looking, for materials, at stiffness when sheared relative to stiffness when stretched, and, for structures, at stiffness when twisted relative to stiffness when bent. That yields ratios that are dimensionless and therefore much less affected by size.

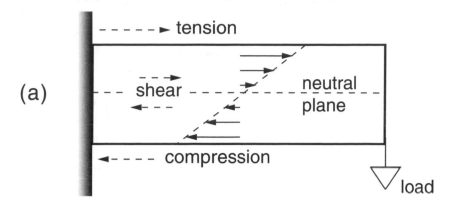

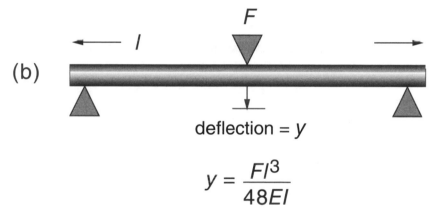

$$y = \frac{Fl^3}{48EI}$$

Figure 11.1. (a) The internal forces that result from a normal force on a beam. (b) A simple apparatus for measuring EI as a composite variable with what is known as three-point bending.

DEFINING THE VARIABLES

We'll begin with three properties of materials and then, to put the materials into structures, two geometric ones. The material properties come from measurements of how something changes shape under an applied force. To shift from a specific sample being tested to a material, one divides force by cross-section to get *stress*. For a tensile test, the complementary variable, *strain*, is extension divided by original length. For shear, strain is the angle turned (as, say, a rectangular solid develops paired parallelogrammatic faces) as a result of a force applied over the face area

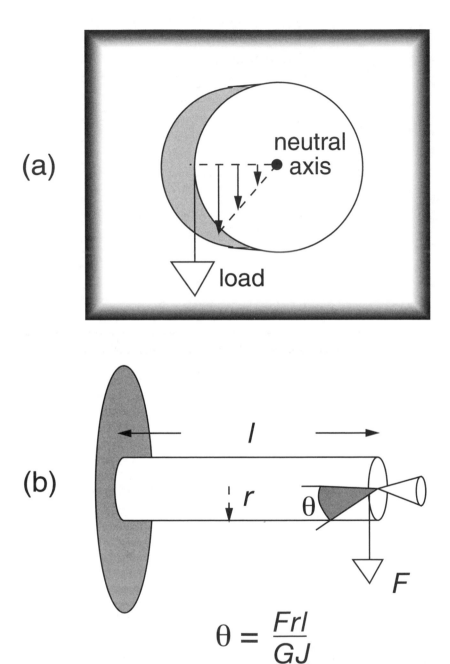

$$\theta = \frac{Frl}{GJ}$$

Figure 11.2. (a) The internal shear forces that result from a torsional load on a beam; compression occurs as well—think of how one squeezes water out of wet fabric by twisting it. (b) An apparatus, perhaps a metal lathe equipped with locking headstock and live center, for measuring GJ as a composite variable.

(rather than cross-section) of a block. Stress has dimensions of force over area; strain is dimensionless.

The *Young's modulus of elasticity*, or tensile modulus, is stress over strain or, better, the slope of a graph of tensile stress (ordinate) against tensile strain (abscissa). Since for biological materials the slope can't be assumed constant (by contrast with metals in particular), one should specify whether a datum gives initial modulus, tangent modulus, or something else. We'll consider initial moduli here, bearing in mind that where stress-strain curves are nonlinear and loads cause severe shape change, we might overlook errors twofold or worse. The *shear modulus* is the equivalent slope for a graph of shear stress against shear strain, again specifying the location on the graph of the particular slope. As angular deformation, shear strain needs no correction for original form. Again, we'll assume initial moduli and keep alert for consequential error.

Stretching an object usually makes it shrink in directions normal to the stretch. But the shrinkage for a given stretch varies from material to material. The common measure of that relationship is *Poisson's ratio*, compressive strain normal to the load relative to the load's tensile strain. Of course the very concept of a single Poisson's ratio for a material presumes isometry, that the material behaves in the same way whatever the direction of the stress. No single ratio can really describe an anisotropic material. That means just about any living or once living material—all but a few are seriously anisotropic. Material can be stretched in any of three orthogonal directions, with compensatory shrinkage in the two other directions. Thus a nicely isotropic metal may have one Poisson's ratio but anisotropic wood or bone will have three times two, or six.

Engineering texts and handbooks cite a formula in which Poisson's ratio, v, sets a relationship between the Young's modulus, E, and the shear modulus, G, of a material:

$$G = \frac{E}{2(1 + v)}. \tag{11.1}$$

Something that retains its original volume (like many biological materials) should have a Poisson's ratio of 0.5 and an E/G ratio of 3.0. I mention the formula and that value to assert their total unreliability for biological materials. The shadows of embedded assumptions, history, and practicality afflict the formulas we borrow from engineers even more strongly than those we get from physicists. Equation (11.1) assumes an isotropic material, one that responds in the same way to loads from any quarter, a condition that the materials of organisms almost never meet. The *inapplicability* of the equation will be crucial in what follows here.

E/G, though, provides a properly dimensionless ratio for comparing several important properties of materials. That last word, *materials*, points up its main drawback. For it to work for structures, all those being compared must have the same shape—although not size. To extend its generality, we need a complementary pair of geometric variables, ones that account for how stresses vary within loaded structures.

When a structure is bent (flexed), both stress and strain vary with distance from the central plane of bending. For a material with a linear stress-vs-strain line (again the usual simplification) both stress and strain distributions increase with loading as in figure 11.1a. Therefore, material contributes to stress resistance in proportion to the square of its distance from a central "neutral" plane. We define our geometric variable as the sum of the squares of the distances (*y*'s) of each element of cross-section from that neutral plane, with each square multiplied by that unit cross-section, *dA*. This *second moment of area* (sometimes, ambiguously, "moment of inertia"), *I*, is thus

$$I = \int y^2 dA. \tag{11.2}$$

For a solid circular cylinder, for instance,

$$I = \frac{\pi r^4}{4}. \tag{11.3}$$

Flexural stiffness (sometimes "flexural rigidity"), the resistance of a structure to bending, is just the product of the material factor, *E*, and the geometric factor, *I*. In practice, it's usually easiest to measure the composite variable, *EI*, as a single operation, as in figure 11.1b.

For twisting loads, material also contributes to stress resistance, now shear stress, in proportion to the square of its distance from the center—again assuming a linearly elastic material. The center, though, is now a neutral axis rather than a neutral plane, as in figure 11.2a. So the corresponding geometric variable, *J*, the *second polar moment of area*, has a nearly identical formula:

$$J = \int r^2 dA. \tag{11.4}$$

For a solid circular cylinder, as before,

$$J = \frac{\pi r^4}{2}. \tag{11.5}$$

Similarly, *torsional stiffness* (or "torsional rigidity"), the resistance of a structure to twisting, is the product of the material factor, *G*, and the

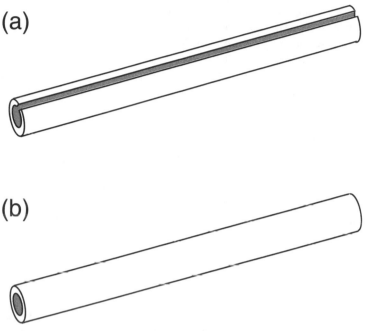

Figure 11.3. (a) An especially twistable structure, one for which conventional calculation of I gives a mechanically unrealistic value. (b) A similar structure that lacks the lengthwise slit and that resists torsion much more strongly—as calculated from I. The pair can be made from a piece of cardboard or plastic pipe for a dramatic illustration of the effect of permitting lengthwise shear.

corresponding geometrical factor, J. Again, measuring the composite variable, GJ, as in figure 11.2b, is usually the simplest procedure.

None of our four key variables is ideally tidy and law-abiding. Either modulus, strictly, works only for one location on a stress-strain graph, as already mentioned. I and J give trouble as well. Figure 11.3a shows a structure, a hollow cylinder with a lengthwise slit, for which calculated J's greatly overestimate measured (and thus functional) GJ's. So measuring the composite variables is not just simpler but will be less likely to mislead.

We now have measures of behavior that, by incorporating I and J, apply to structures rather than merely to materials. Expressing them in a ratio gives what we sought at the start, a size-correcting dimensionless basis for comparisons. That ratio of torsional to flexural stiffness, then, is GJ/EI. I prefer to take an arbitrary further step and shift to its reciprocal, flexural to torsional stiffness, or EI/GJ.

That inversion moves us from the world of the engineer to that of the biologist. A technology that values rigidity represents its variables as resistance to deformation, hence flexural and torsional *stiffnesses*. GJ/EI thus gives twist resistance relative to bend resistance. A biologist looking at natural design, where achieving rigidity seems less often the primary goal, does better with ease of twisting relative to ease of bending—a shift in thinking from terms of stiffnesses to ones of compliances. That means EI/GJ, in conventional terms a "twistiness-to-bendiness" ratio. As we'll see, EI/GJ has the additional advantage of yielding values most often above 1.0. Etnier (2003) has defined a "stiffness mechanospace" for elongate biological structures with just this ratio.

Combining the value of 3.0 given earlier for the E/G of a circular cylinder of an isotropic, isovolumetric material with equations (11.3) and (11.5) for I and J provides baseline values for this twistiness-to-bendiness ratio. For such a cylinder, EI/GJ will be 1.50. The corresponding values for square and equilaterally triangular sections are 1.77 and 2.49. Even when comparing cylindrical structures, where E/G would do, I'll most often cite values of EI/GJ—mainly to anticipate discussion further along of shape effects. For consistency where data exist only for the moduli, I'll assume a circular cylinder, so $EI/GJ = E/2G$. Table 11.1 collects the large number of values of the EI/GJ ratio that will be cited.

Not only should the ratio of flexural to torsional stiffness, our twistiness-to-bendiness ratio, help assess the role of torsion, but it may provide a simple index of the degree of functional anisometry of a material or a structure. We'll return to this second role near the end of the chapter.

For Example, Metal versus Wood

By adopting the formulations of the mechanical engineers, we can take their tabulated data as context for a look at biological structures. Their classic structural materials—although rarely used in quantity before the nineteenth century—are, of course, metals. The introduction just given comes from that era; assumptions such as that of linear stress-strain plots (and thus strain-independent moduli) retain the odor of their metallocentric world.

Data for both flexural and torsional stiffness abound, the former critical for building large structures, the latter important for choosing rotating shaftwork in order to transmit power. E/G values range between about 2.4 and 2.8, corresponding to Poisson's ratios of 0.2 to 0.4, significantly below that isovolumetric 0.5. When pulled upon, metals turn out to expand volumetrically as they extend. We just use them at such low strains that we can usually ignore the volume change. For our paradigmatic circular cylinders, that range of E/G-values gives twistiness-to-bendiness

TABLE 11.1

Representative values of the twistiness-to-bendiness ratio, *EI/GJ*. Parentheses give number of species from which values are averaged. Details, additional values, and references are in the text.

Circular and assumed circular structures	
Isotropic, isovolumetric cylinder	1.50
Steel shaft	1.3
Commerical (dry) wood (99)	7.15
Tree trunks (5)	7.34
Mature woody vines (5)	3.13
Woody roots (1)	2.34
Long bones (femurs) (2)	2.86
Primate mandibles (2)	1.55
Circular petioles (1)	2.8
Gorgonian corals (13)	3.9
Jointed beams (5)	4.3
Sunflower shoot (1)	1.4
Structures with noncircular cross-sections	
Grooved or flat petioles (3)	5.9
Daffodil stems (1)	13.3
Sedge stem (1)	~36
Banana petiole (1)	~75
Locust hind tibia (1)	6.4
Feather shaft (1)	4.8

ratios, *EI/GJ*s, of 1.2 to 1.4. By altering shape, one can't easily push the values downward, but they can be elevated without limit by using ever less cylindrical sections—cross shapes, I-shapes, and so forth.

As important as metals and important to humans far earlier is wood—although as dried and shaped pieces of tree rather than the living biomaterial. Wood is anything but isotropic. Every woodworker must pay constant attention to the direction of the grain, critical to its mechanical behavior. We tame the anisotropy by gluing pieces broadside with crossed grain, forming sheets of plywood and particle board that behave similarly in at least two directions. That anisotropy gives it ratios of E to G far higher than those of metals. For circular cylinders, I calculate an average EI/GJ ratio of 7.15, not 1.4, for the ninety-nine kinds of wood tabulated by Bodig and Jayne (1982). Of course one has to specify direction for any anisotropic material—here E is for a longitudinal pull and G for a longitudinal-radial plane. Despite conspicuous differences in microstructure and striking variation in practical performance, the ratio varies little from wood to wood, with a standard deviation of 1.21—16.9 percent of the mean.

The ratio varies less than the density of the woods, the latter with a standard deviation of 21.1 percent of the mean value. It also varies less than both E (20.5 percent) and G (20.8 percent), which is to say that the two covary and that equation (11.1) still casts its shadow. Still, applying that equation as given produces a Poisson's ratio of 12.3, which would imply a fabulous radial shrinkage for a modest longitudinal tension.

Textbooks on materials and mechanical design, even older ones, appear silent about the practical consequence of the great difference between E/G (or EI/GJ) of metals and of woods or about the high ratios for woods. Silence in an application-driven field suggests minimal importance. But one can at least envision points of relevance. The high values for woods should affect the behavior of unipodal wooden furniture. Thus wooden lecterns and pedestal tables, even if adequately resistant to bending, will be relatively prone to twisting. But one suspects that anticipation of such problems comes more from experience than calculation.

CYLINDRICAL STRUCTURES

Tree trunks. Most tree trunks have circular cross-sections, so geometric issues can be put aside, and EI/GJ can be equated (recalling equations 11.3 and 11.5) with half of E/G. That makes tree trunks an obvious starting point for asking what our ratio might tell us. First, though, do those data just cited for wood, however relevant to its role as construction material, say anything about wood as the material of a tree—as opposed to its performance as a sliced and dried commercial material?

Common experience tells us that dead wood and live wood differ mechanically—a dead twig snaps; a live one bends. Unsurprisingly, measurements confirm the observation. Hoffmann et al. (2003) reported direct comparisons between dried (at 55 percent relative humidity) and rehydrated sections of the stems of several tropical lianas (woody vines). Dried *Bauhinia* stems had fully twice and dried *Condylocarpon* stems

1.4 times the Young's moduli of rehydrated stems. Ratios of shear moduli for the two were essentially the same, 1.9 and 1.4. Of especial present interest, the quotients, E/G or EI/GJ, thus differed only minimally.

The same result emerges when my values (Vogel 1995b) for freshly collected trunks of small trees are compared with those of Bodig and Jayne (1982) for lumber of the same species. For the five species in common, two gymnosperms and five angiosperms, slicing and drying raises E by factors of 1.8 and G by 1.75 on average, with lots of variation and no obvious interspecific pattern. So EI/GJ changes little—it drops an insignificant 2.4 percent. In short, while drying changes the moduli considerably, it has curiously little effect on the ratio of the moduli. Put another way, whatever gives wood its peculiarly high E/G ratios seems not to depend on its water content. Trees certainly aren't hydrostatically supported systems.

What, then, about the high values of our twistiness-to-bendiness ratio in living trees? While data for Young's modulus are common enough, shear modulus seems only rarely to be determined. My own measurements of the two moduli on freshly cut lengths of a few trunks suggest a general pattern. For three hardwood trunks, EI/GJ averages 8.7, with little variation; Bodig and Jayne (1982) give 8.0 for prepared wood. Bamboo culms (not "trees" in the strict botanical sense) give an average value of 8.6, insignificantly different. Two softwoods, a loblolly pine and a red cedar, have lower ratios, 6.1 and 4.4. (Bodig and Jayne give 8.2 and 4.1.) The differences between hardwoods and softwoods, whatever the present values, should not be taken as general, judging from extensive data (Bodig and Jayne 1982, again) on prepared wood, where EI/GJ averages 6.91 ($s=1.12$) for fifty-two types of hardwood and 7.51 ($s=1.13$) for forty-seven types of softwood.

Remarkably, the EI/GJ ratios for both prepared wood and freshly cut lengths of small trunks far exceed the 1.50 of isotropic, isovolumetric circular cylinders. Do high ratios hold functional significance; might they represent a direct product of natural selection? Any assertion of direct selective significance requires stronger evidence. One might well be viewing some indirect consequence of the design of xylem as sap conduits. After all, a set of parallel, longitudinally oriented pipes should be anisotropic. Put rubber bands around a bundle of thin, cylindrical, dry pasta, and the structure will twist more easily than it bends, at least if the strands are not too strongly squeezed together. Still, one wonders . . .

Unless especially symmetrical in both shape and exposure, a tree in a wind will feel both bending and turning moments. Accommodating that turning moment with some twist might lessen the associated bending moment via drag-reducing reconfiguration. The softwoods with low values of EI/GJ, loblolly pine and red cedar, are relatively narrow-crowned, symmetrical trees, less subject to torsional loading than most others. But we should resist the seductive appeal of facile functional rationalization. Thus I find the trunks of trees with woven rather than straight-grained wood, such as sweetgum, sourwood, and sycamore, especially hard to split lengthwise for firewood. Yet the ratios for such trees don't differ much from ones that split easily, such as oaks and tulip poplar. That's also a reminder that we're asking whether a twistiness-to-bendiness ratio is a relevant structural variable, not whether it's the only relevant one. It doesn't appear to covary with failure point, whether expressing the latter as strength, extensibility, or work of fracture. And it's unlikely to correlate with energy absorption before failure.

Woody vines. Of particular present interest are recent measurements on woody vines, or lianas, common and diverse in tropical forests. The woods of mature lianas mainly bear straightforward tensile loads rather than the more complex compressive, flexural, and torsional mixes of self-supported plant axes. Concomitantly, climbing members of most lineages (and lianas have evolved many times) have lower dry densities and much wider vessels than do the self-supporting trunks considered so far. I looked at the attached, climbing stems of two woody vines, a native grape, *Vitis rotundifolia*, and an introduced and escaped ornamental, *Wisteria sinensis*; the first gave $EI/GJ=2.66$, the second an average of 4.48 with especially wide specimen-to-specimen variation (Vogel 1995b). These values lie well below the ratios for almost all tree trunks.

But most such plants support themselves at an early stage, and the apical regions continue to do so as they reach out for external assistance. Early stages and the apical portions of climbers have flexural stiffnesses typical of woody trunks; stiffnesses of later stages are up to an order of magnitude lower. Monocots, which lack secondary (that is, radial) growth, provide most of the exceptions (Rowe et al. 2006). The fewer data on torsional stiffness don't seem to drop in the same manner. For climbing specimens of *Croton nuntians*, EI/GJ drops during ontogeny from about 9 to about 0.8 (Gallenmüller et al. 2004). For *Bauhinia guianensis* it drops from 9.5 to 5.9, while for *Condylocarpon guianense* it goes from the exceptionally low 2.6 to a still lower 1.8 (Hoffmann et al. 2003). In short, after giving up self-support, all climbers so far tested have especially low EI/GJ ratios.

As noted by Gallenmüller et al. (2004), the ontogenetic change makes functional sense. Not only must freestanding plants support themselves (high E), but climbing plants may gain more from flexibility in bending (low E) beyond mere material economy. Flexibility should limit loading caused by movements of their supportive hosts. But equivalent ontogenetic change for shear modulus (G) should yield little benefit, because high torsional flexibility shouldn't be particularly disadvantageous at any stage.

(Having given data for low EI/GJ ratios for mature lianas from several studies and made a case for their functional significance, I don't know how to view a study by Putz and Holbrook 1991 that reports E and G values for twelve tropical lianas from Puerto Rico. Assuming circular sections (no data are given), they give ratios ranging from 2.0 to 52.8 with an average of 10.0. Omission of the high outlier reduces that average to 6.09. For five trees, by comparison, they give a range of 3.1 to 33.5 and an average of 13.2; again omitting the high outlier drops the latter to 8.1, similar to my data. They provide no information on stage or habit of the specimens.)

Roots. Despite the obvious mechanical importance of roots, data are scarce. One guesses that the most serious loads are tensile. One guesses further that a very high Young's modulus could be disadvantageous, quite unlike the situation for an upright trunk—a little "give" will lower the peak stresses of impulsive loads and will facilitate load-sharing among an array of nearly parallel roots. And one anticipates less functional relevance for shear modulus. So the EI/GJ ratio may mainly illustrate the possible variation in mechanical behavior of woody structures that have lengthwise conduits and thus anisotropic "grain."

To provide such a contrast, I dug up and tested a few roots (Vogel 1995b). Young's moduli are indeed low. For instance, for loblolly pines (*Pinus taeda*), $E=6.01$ gigapascals for trunks but only 1.13 gigapascals for roots, a difference that's obvious when handling pieces. Shear moduli differ less, with EI/GJ dropping from 6.12 to 2.34. So roots are much less anisotropic than trunks, parallel conduits notwithstanding.

Bones. The nearest analog to tree trunks among animals must be the long leg bones of large terrestrial mammals—elongate, vertical, gravitationally loaded, and cylindrical or nearly so. (Most but not all long bones are close to circular in section. Cubo and Casinos 1998 give extensive data for those of birds and mammals, noting as an extreme the cross-sectional aspect ratio of 2.0 of the tarsometatarsi of a parrot.) Long leg bones must face more diverse loads than do tree trunks, such as those from the various muscles and postures and from the impact loads of running. But they won't often feel major torsional loading—except when we attach long, transverse levers, skis, with indequate provision for load-sensitive release. In addition, they lack stiff-walled, lengthwise, vascular elements like xylem.

Long bones are stiff, with Young's moduli, about 20 gigapascals, about three times greater than those of tree trunks and exceeding even bamboo culms, about 15 gigapascals. But they're far less anisotropic by comparative measurements along the principal axes. Or in terms of our ratio of EI/GJ, our human femurs average 2.59 and bovine femurs have a similar value of 3.14 (Reilly and Burstein 1975).

What about bones of more complex shape and ones that might bear more diverse loads? Dental interest, of both the extant and the extinct, has stimulated measurements on primate mandibles. Those of both rhesus macaques and humans have somewhat higher Young's moduli than those of corresponding femurs, but they have notably lower twistiness-to-bendiness ratios—1.48 for macaques (Dechow and Hylander 2000) and 1.62 for humans (Schwartz-Dabney and Dechow 2003)—applying I- and J-values for cylinders for the sake of comparison. These ratios do not quite correspond to those of isotropic, isovolumetric materials, though. Poisson's ratios for similarly stressed bone run between about 0.2 and 0.4, significantly non-isovolumetric. Equation (11.1) and a value of $v=0.3$ (and assuming a circular cross-section) implies a still lower ratio, 1.3, were they isotropic. So ratios for bones of any kind (as far as we know) are at or below ratios for woods. And the twistiness-to-bendiness ratio appears to give a rough but convenient indicator of anisotropy.

Other items with circular cross-sections will appear further along, deferred to facilitate comparisons with specific noncircular ones.

Noncircular Structures

So far (except a comment or two on mandibles) we've looked only at elongate circular cylinders, in effect holding shape constant. My colleague Steve Wainwright (1988) argues that such cylinders form a morphological baseline for a diversity of macroscopic biological forms. So I might claim something beyond heuristic justification for that starting point. But we now move on, recalling that deviation from circular sections increases torsional flexibility relative to flexibility in bending. And, as we've just seen, incorporation of anisotropic materials can cause analogous increases in EI/GJ. How might nature combine these two routes, geometric and material, to the same end?

Petioles and herbaceous stems. My original impetus for invoking the ratio came from a look at how the leaves of a variety of broad-leafed plants responded to high winds (Vogel 1989). Most reconfigured into cones and (for pinnately compound ones) cylinders. That reduced flutter, which might shred them as it does flimsy flags in winds, and drag, which might uproot or break the parent trees. When leaves were exposed individually, petioles (leaf stems) appeared mainly loaded in tension. But when, as should be more common in nature, groups of leaves were exposed to winds, most reoriented into stable, low drag clumps. Clumping requires that petioles twist lengthwise. So petioles, acting as cantilever beams that extend leaf blades, must resist bending, but at the same time they should accommodate twisting. That argues for elevated EI/GJ ratios, whether achieved by geometric or material specialization.

I found that petioles indeed had EI/GJ values well above the isotropic, isovolumetric baseline—whether or not their cross-sections were circular. Still, noncircular ones had higher ratios than did circular ones. A typical circular petiole, red maple (*Acer rubrum*) had a ratio of 2.8; ones with some lengthwise grooving (as in plate 11.1 left) averaged about 5.0. One might guess that grooving should be more common among shorter than among longer petioles—short ones would need more twist per unit length for a given blade rotation. Mami Taniuchi (unpublished) examined the literature on petiole lengths and cross-sections; she found some indication of such a correlation, but at the margin of statistical significance. While we can't safely assert specific adaptation—that structural variation has been driven by functional imperatives—convergence would provide some evidence that natural selection operated on this specific feature (Vogel 1996b).

Bilateral flattening (rather than grooving) characterizes the genus *Populus*, which includes trees such as cottonwood (*P. deltoides*) and quaking aspen (*P. tremuloides*). My value for bilaterally flattened petioles of white poplar (*P. alba*), bent laterally (as in pictures of wind-driven clustering) was 7.7. Niklas (1991) gathered more extensive data for *P. tremuloides*; his petioles increased in EI/GJ from 2.11 to 9.62 as they developed. All these species have a strong propensity to oscillate in modest winds (the mechanism was described by Bschorr 1991), and the possible function of that visually and aurally attractive habit has stimulated a lot of speculation. I once put a variety of individual leaves in the strongest and most turbulent wind available, about 30 meters per second. While most leaves shredded, those of *P. alba* suffered no obvious damage. From that and casual observations through binoculars of leaves in thunderstorms, I suggest that instability, especially torsional, at modest speeds goes along with good reconfigurational ability at high speeds. *Populus* leaves, with their flattened petioles, are just the extreme case of both. In short, the low-speed shimmering is nothing more than an incidental concomitant of good high-speed performance. That's reasonable enough for the trees of a genus especially common in windy places—high altitudes, open plains, and coasts.

Several cultivated herbaceous stems gave analogous results. For tomatoes, with circular stems, $EI/GJ = 3.9$; for cucumbers, with cruciform-shaped stems, $EI/GJ = 5.4$. While the former grow upright and freestanding, the latter are either recumbent or climbing—but this limited comparison can only hint at functional significance.

Another comparative study also found relatively low values for circular sections, but values that still remain above the baseline expectation for structures made of isotropic materials with reasonable Poisson's ratios. Niklas (1997a) looked at the hollow internodes of six herbaceous species that had circular sections between their crosswise septae. Despite wide phylogenetic diversity and over fourfold ranges of E's and G's, the ratios of $E/2G$, or EI/GJ, varied very little from 2.5.

However far beyond our baseline, even the highest figures mentioned so far are far below the extremes. The flower stems of daffodils (*Narcissus pseudonarcissus*) are hollow and lenticular rather than circular in section. They bear single apical flowers well to the sides of their long axes, flowers that "dance" in the intermittent gusts common near ground level—as alluded to by several British and American poets. As put by William Paley (1802), posthumously famous as the darling of the anti-Darwinians, "All the blossoms turn their backs to the wind, whenever the gale blows hard enough to endanger their delicate parts." As confirmed by Etnier and Vogel (2000), wind on the off-axis flowers loads their stems torsionally, causing them to swing around and "face" downstream. That reduces the drag of the flowers, in effect the flexural load of the wind, by up to a third. We reported an average EI/GJ value for daffodil stems of 13.3, with the remarkably low standard deviation of 1.0. By comparison, the stems of tulips, with circular stems that bear axially symmetric flowers, gave EI/GJ values of 8. 3±3.2.

Two other structures have values far higher still, the flower stems of a sedge (Ennos 1993b) and the petioles of banana leaves (Ennos et al. 2000). Sedge (*Carex*) flower stems stand erect but curve to one side. Thus, winds will load them both flexurally and torsionally. They're triangular in cross-section; for an isotropic, isovolumetric material that might raise EI/GJ from 1.50 to 2.49. In fact, their ratios range from 22 to 51. While both flexural and torsional stiffness drop by more than an order of magnitude with height above the ground, EI/GJ changes much less, peaking about halfway up. These radically high values most likely come from a ring of mechanically isolated strips of lignified material, each running up and down the stem's periphery.

Similarly, banana petioles, exceptionally large for herbaceous structures, extend both upward and outward. Unlike sedge flower stems, they have U-shaped sections. But like sedge stems, they have peripherally concentrated, longitudinal, isolated, lignified elements. These play a major role in producing twistiness-to-bendiness ratios from 40 to 100, the highest yet measured in a natural structure. These high EI/GJ values ensure that rather than bending, banana petioles will twist away from the direction of the wind.

Other noncircular structures. Most of our data on elevation of twistiness-to-bendiness ratios with noncircular sections come from stems and petioles. But that predominance need not indicate an unusual reliance on the device by such structures. It's more likely no more than the predilections of investigators and a rare case where plants have drawn more attention than animals. Maybe the limited animal data will at least animate further work.

Gorgonian corals, common in shallow tropical seas, are erect but relatively nonrigid. Jeyasuria and Lewis (1987) reported E and G values for thirteen species from the West Indies, although with no information on cross-sectional shape except a note that some were circular, others elliptical. Assuming circularity produces EI/GJ values from 1.6 to 6.5 with an average of 3.9. Of interest here is a comment that species in which polyps surround the circumference of the branches twist more easily, while those with single rows of polyps on the sides of the branches twist less easily. They interpreted the difference in terms of a greater need to maintain torsional orientation where polyps are aligned in rows.

The joints of arthropod legs rarely if ever incorporate analogs of our hips, ankles, shoulders, and wrists, joints capable of considerable rotation. When dismembering a decapod crustacean, for instance, one quickly learns that legs disarticulate when twisted. Perhaps absorbing torsional loads through shaft twisting could reduce the loads on such torsionally vulnerable joints. Most leg segments of arthropods look circular, but a strip of especially thin cuticle often extends lengthwise. This thin region may be most familiar in the walking legs (pleopods) of crabs and lobsters, evident whether they're fresh or cooked. It might provide the equivalent of a groove or other I-lowering device, increasing the torsional flexibility of the segments.

Despite a wealth of descriptive information and illustrations, I found no measurements of the behavior of torsionally loaded leg segments. So in parallel to the measurements on petioles, I tested a few hind tibias of freshly caught locusts (probably *Dissosteira carolina*), these being the least tapered segments of their most powerful jumping legs. EI/GJ averaged 6.4, providing some support for the argument. Specimens were quite vulnerable to local buckling, so only very slight torsional strains could be imposed, and three-point bending tests had to be done with loops of thread rather than the usual point contacts.

The vane-bearing shafts (rachises) of the long, outer feathers of birds, especially those of tails and wings, are more obviously noncircular in section. They may have lengthwise grooves on their lower surfaces (as in plate 11.1 middle), or they may be nearly square in section with thinner lower than upper sides. I made a few measurements on pieces of shaft from the primary wing feathers of song sparrows (*Melospiza melodia*), obtaining EI/GJ-values of about 4.8—again indicative of a structure that preferentially accepts torsion.

One can make a similar argument, here based on flight aerodynamics. Wing feathers, like propeller blades, should not bend excessively, so function demands high flexural stiffness. At the same time, animal wings, incapable of full rotation, must alternately move up and down. So angles of attack of wings and primary (wing tip) feathers for producing lift and thrust must shift between half-strokes. For primary feathers that means reversing their lengthwise twist. Adequate torsional flexibility can enlist aerodynamic forces for that switch in twist, avoiding dependence on muscles and nerves. Thus, the feathers ought to have high EI's and low GJ's. That the grooving or thinning is ventral rather than, as in petioles, dorsal (=abaxial), makes sense as well. While petioles hang from branches, birds hang from feathers, and feathers in flight mainly bend upward rather than downward—although a bit of downward bending might happen during upstrokes. (One can see the shift in loading by watching the wing of an airplane through an adjacent window as it leaves the ground.)

Another set of noncircular structures worth investigating are the neural (dorsal) spines sticking up and rearward from the centra of the vertebrae of large, bony fishes. These (as in plate 11.1 right) often have U-shaped sections. Any resulting increase in torsional flexibility might be important when a fish bends its body. If the spines were fully vertical, bending would not impose a torsional load, but their rearward tilt requires that they twist as the overall fish bends.

PLANAR SYSTEMS

So far we've looked at cylindrical or nearly cylindrical structures. Hollow structures were fully enclosed, with no lengthwise openings into their lumens. Planar and near-planar structures experience and respond to torsional stresses in ways of equal biological cogency. Among nature's designs, flat surfaces are simply less common—or at least less diverse. Not that they're truly rare—examples include the leaves of higher plants, many macroalgal fronds, the vanes of feathers, and the wings of insects.

Most insects use indirectly acting muscles to power the strokes of their flapping wings, muscles that attach at neither end to the wing articulations but instead act by reshaping their thoracic cuticle. The small direct muscles that insert on the bases of the wings supposedly rotate and camber the wings, making the changes necessary between each alternating half-stroke. In addition, they've been held responsible for the reversal of lengthwise twist between half-strokes needed (as mentioned for the wing feathers of birds) to maintain a near-uniform angle of attack along a wing's length. The precise phasing of these direct muscles drew little attention despite the severe neuromuscular demands of their role where wings beat hundreds of times each second.

Ennos (1988) directed attention to a more realistic mechanism, one in which their intrinsic and locally tuned torsional flexibility enabled insect wings to use aerodynamic forces for these rapid changes in wing contour. That paper forced reevaluation both of the role of the direct muscles and of the role and arrangements of wing veins. A second paper (Ennos 1995) provided a general analysis of the torsional behavior of cambered plates, as found in leaves (especially grasses), feather vanes (as noted earlier), and cuticular plates in arthropods.

Coincidentally, the first successful human-built aircraft took advantage of torsional flexibility in much the same manner—if without the rapid reversals demanded by flapping. In the 1903 Wright Flyer, the pilot controlled turns by shifting a slide that, through cables, twisted opposite wings so they would produce different amounts of lift. Vertical struts connected upper and lower wings, but the wings lack the cross-bracing that would normally provide torsional stiffness. What look like cross-bracing (and are inaccurately shown as such in many drawings) were, in fact, those wing-twisting cables. Hinged ailerons, as still used, soon replaced wing warping, initially as Henri Farman's attempt, in 1908, to circumvent the Wrights' patent on their system of control (Anderson 1997).

Rolling a flat surface into a cylinder without sealing the joint to prevent shear produces a cylindrical beam or column that resists bending but has high torsional flexibility. For a model one need only roll up a sheet of paper. The arrangement characterizes many xeromorphic (drought-adapted) grass blades. The usual explanation has the device reducing water loss; noting that stomata are on the inside of the cylinder, it views the roll as a functional addition to the sunken stomata of more planar xeromorphs. That the roll might also have a mechanical role seems not to have been considered. The tops of tall grasses ordinarily bend to one side, so wind will load blades torsionally. As suggested earlier, torsional flexibility can reorient such a structure so more surface area is parallel to flow and downwind from other surfaces. That would decrease drag and reduce the chance of flexural buckling—in crops this kind of damage is called "lodging."

The parent phenomenon, aeroelasticity, has been of considerable interest to aircraft designers, but they more often focus on trouble rather than on utility. Tilting a wing to change its lift usually moves the center of pressure fore or aft, changing the torque on the wing and on its attachment to the fuselage. In at least one aircraft used during World War I, a Fokker D8, the wings sometimes detached as a disastrous result (Gordon 1978). But aeroelasticity has occasionally played a positive role in aircraft. For example, a small, high-performance, military jet (a US F-18) has been fitted with torsionally aeroelastic wings. It deserves more attention from biologists. We touched on it when considering the reconfigurations of leaves in winds; it takes on similar importance for tree trunks and similar structures in rapid flows (Miller 2005); and it might be used to induce oscillations that detach seeds and spores into dispersing currents of air or water.

Jointed Systems

Joints and their behavior in torsion came up in connection with arthropod legs. Joints ordinarily bring into the picture muscles and short-term adjustments of mechanical behavior. We can, for instance, deliberately stabilize our wrists so they resist either twisting or bending as a particular task requires. When we jump, we spontaneously adjust the compressive stiffness of our legs, mainly at our ankles, so the stiffness of legs plus surface remains nearly constant over a wide range of surface compliances (Ferris and Farley 1997). We also adjust ankle torsional stiffness, if slightly less so (Farley et al. 1998). Whether our actions are voluntary or involuntary and whether for wrists or ankles, we can independently control the two stiffnesses.

Several biological systems use beams consisting of alternating joints and stiff portions. Etnier (2001) looked at such multi-jointed systems in horsetails (*Equisetum*), crinoid (echinoderm) arms, and crustacean antennae, obtaining EI/GJ values not greatly different (1.8 to 6.6) from those of simple biological beams and columns. Here, though, functional significance for values above that baseline of 1.5 for isotropic, isovolumetric cylinders remains unclear—no pattern in the behavior of such jointed systems has yet emerged. What has emerged is further evidence that natural structures can achieve a range of values with joints and muscles as well as by controlling the material and geometric characteristics of elongate, passive, solid elements.

Dissecting the Variables

For simple circular cylinders, again, E/G would serve just as well as EI/GJ for a handy index for the degree of anisotropy. Admitting a role for shape, as defined by the two moments of inertia, permits us to dissect anisotropy into an instructive trio of components. Stretching the usual meaning of anisotropy we can regard the ratio of the elastic moduli, E/G, or as used here, $E/2G$, as "material anisotropy," with (for the latter) a baseline of 1.5 for isovolumetric cases. That of the second moments, $2I/J$, then becomes "geometrical anisotropy," with a baseline of 1.0 for cylindrical sections. And their product, EI/GJ, constitutes "structural anisometry," in the form of a twistiness-to-bendiness ratio.

From any pair of ratios one can get the third. The easiest route will usually consist of measuring EI/GJ, obtaining (with the caution previously noted) $2I/J$ from cross-sectional shape, and then calculating $E/2G$. For only a few cases can EI/GJ yet be teased apart in this way. For daffodil stems, for instance, $E/2G = 10.0$ since $EI/GJ = 13$ and $2I/J = 1.3$ (Etnier and Vogel 2000). For sedges (Ennos 1993b) a $2I/J$ of 1.25 implies an $E/2G$ of roughly 30. Together with the elevated EI/GJ ratios of circular beams such as bones and tree trunks, such values suggest a general characteristic of biological designs. Structural anisometry comes far more from an unusually high E relative to G, that is, from high material anisometry, than from high I relative to J, that is, from geometric anisometry. Put another way, deviation from circularity commonly indicates a high EI/GJ, but it usually represents the lesser causative element. For this reason EI/GJ provides almost as good an indicator of material anisometry as $E/2G$.

That contrasts strikingly with the structures we humans make. We take a piece (or melt) of metal or plastic and then treat it as a homogeneous material in fabricating some desired shape. In effect, we accept as given a value of $E/2G$ of about 1.4. Particularly in large structures, most shifts in the structural ratio, intentional or incidental, come from adjustment of geometric anisometry. Structures such as I-beams and metal fence posts get their high EI/GJ ratios entirely as consequences of their cross-sectional shapes. And the factories of our mechanized, mass-productive society leave to the province of craft workers the paradigmatic high-$E/2G$ material, wood.

LOOKING STILL FURTHER AFIELD

Hydroskeletons. Everything so far implies, first, that material and structural specialization can raise the twistiness-to-bendiness ratio above a baseline value and, second, that functional advantage may be gained by its elevation. Extending the ratio downward puts a different set of biological designs on the same scale. Lengthwise anti-grooves, that is, ridges, won't lower EI/GJ, but a widespread arrangement will do so quite effectively. In it, an incompressible but highly nonrigid core is surrounded by a flexible skin, the latter with fibers running helically in both directions. These so-called hydroskeletons provide support, among others, for limp annelid and stiff nematode worms, for the tube feet of echinoderms, for the mantles of squids, and for the bodies of sharks.

Twisting such a fiber-wound, pressurized cylinder one way puts tensile stresses on one set of fibers; twisting it the other way stresses the other set. In all such systems the fibers in the outer membrane are relatively inextensible, so they strongly resist such torsion. At the same time, little except stretch of the membrane itself limits bending. The net effect will be a low twistiness-to-bendiness ratio. An extensive literature on hydroskeletons rarely mentions their torsional stiffness except parenthetically. Clark and Cowey (1958) provide the classic description, Alexander (1988a) puts their operation into the context of motility as well as support, and Niklas (1992) extends the discussion to plants. (The only known hydrostatic supportive systems with lengthwise and circular reinforcement rather than such crossed-helical windings are those of nonbony mammalian penises, as discovered by Kelly 2002. Their function demands resistance to lengthwise compression, provided by their combination of circumferential fibers and a constant volume interior. Who cares that erect penises lack torsional stiffness?)

Unfortunately, hydrostatic supportive cylinders do not lend themselves to the kind of mechanical testing that can determine EI/GJ ratios—imagine an apparatus for twisting a worm. I made a few measurements (Vogel 1992) on a partially hydrostatic system, young shoots of sunflowers (*Helianthus*) about 10 to 15 millimeters high—the average ratio was 1.4, slightly below our baseline of 1.5. But the scatter was wide, with some specimens yielding values around 1.0. Fully hydrostatic structures should have even lower values.

Hydroskeletons occur mainly in aquatic organisms, perhaps no more than a matter of water's availability. But perhaps the requirement that mature terrestrial systems hold heavy structures erect is a further factor behind the relative scarcity of classic hydroskeletons in such systems. After all, for typical terrestrial, gravitationally loaded systems, high EI/GJ ratios should be advantageous and low ratios maladaptive.

A tale of a tall building. Finally, a case in which this twistiness-to-bendiness ratio—and torsion generally—turned out to be unexpectedly relevant. This tale of a faulty tower comes from Levy and Salvadori (2002). When completed, in 1974, the John Hancock Tower, a tall (234 meters), slender building, in Boston, Massachusetts (plate 11.2 left), was applauded for its elegance. Several problems, though, delayed its occupancy for several years. Among the effects of wind, falling exterior glass panels gained the most notoriety, but a more interesting failing was the unpleasant, wind-induced, twisting motion of the building. Between the high aspect ratio of its cross-section, about 3:1, its rhomboidal shape, and the much-admired lengthwise grooves running up the smaller two sides, the structure turned out to be unexpectedly lacking in torsional stiffness—too little GJ for its EI.

The cure, not inexpensive, consisted of two parts. About 1500 tons of diagonal bracing were added. And, occupying the far ends of the next-to-the-top floor and resting on a thin layer of oil, a pair of 300-ton masses were connected to the structure through springs and shock absorbers (plate 11.2 right). These passive dampers, tuned to oscillate in opposite phase with the building, compensated for most of the torsional motion. (For security reasons, visitors can no longer view the damping system. An omnidirectional analog, part of the original design, can be seen in Taipei 101, in Taiwan.)

Nature's designs may be less rigid than those things we build, but perhaps hers can provide guidance if one applies an adjustment of scale. Traditional rigidity takes a disproportionate amount of material in very large structures. Our tall buildings sway and our large aircraft flex their wings. So natural design at the more modest scales of animals and plants may hold relevance for our efforts when we work at much larger scales.

CHAPTER 12

Keeping Up Upward and Down Downward

When You're Perturbed . . .

In defining an organism's immediate physical situation, one begins with position and orientation. Position always matters, even if no exact specification need always be given for the pelagic or the aerial. And only in a few instances can orientation be ignored—for instance for non-motile spherical unicells in a continuous medium, perhaps a colonial *Volvox* (plate 1.1) in a pond, perhaps under some circumstances spherical eggs and nuts.

We ordinarily treat orientation mechanisms as matters of coordination, topics in neurobiology. Detectors, in particular proprioceptors, provide information based on which nervous systems direct appropriate muscular activity. We're less often concerned with underlying physical situations, with potential perturbing forces, with options available for reorientation, and with devices for maintaining orientation. (But I must immediately applaud a symposium held a few years ago, reported as Fish and Full 2002.) One chapter certainly can't do the subject justice—especially since we know quite a lot about the subject in a sometimes scattershot way—but perhaps the main bases can be touched.

Elementary physics texts recognize three mutually exclusive situations, ones in which bodies subject to a gravitational (or analogous) force are either stable, unstable, or neutral in position. As in figure 12.1, the distinctions hinge on differences either in their own mass distributions relative to their footprints or on the contours of the supporting surfaces. (Substratum contour will be put aside as a second-order issue, left to people concerned with behavior.)

A neutrally stable object simply has no preferred orientation. At most, perturbation adds rotational momentum, which then keeps the thing rotating. The commonest neutrally stable objects are rigid cylinders (one neutral axis) and spheres (two such axes) lying on rigid surfaces, and objects in continuous media whose centers of gravity coincide with that of the fluid they displace. I'll say no more about such cases, relatively uncommon among macroscopic systems, just noting that neutral stability opens a possibility for ground-level wind dispersal—one exploited by, at least, Russian thistles (tumbleweeds) in the drier parts of North America.

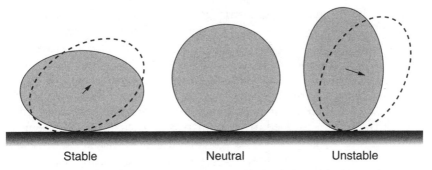

Figure 12.1. The stability of three gravitationally loaded rigid bodies on horizontal, rigid substrata—assuming, of course, uniform density and no other forces.

Almost all our passive possessions are stable around at least two axes—tip one a bit north or south, east or west, and it returns by itself to its previous orientation. They're stable because the work of perturbation raises the object's center of gravity, and a gravitational restoring force then drives its subsequent reorientation. Put another way, the perturbation generates a restoring force, and the system enjoys the resulting negative feedback. For instance, many pencils have hexagonal cross-sections to impart a bit of stability to their rotational axes. Petroski (1990) traces the origin of the practice, one especially handy for cultures with sloping desks. By contrast, while an unstable object may be stationary, any perturbation will upset its balance and produce runaway reorientation—a process with positive rather than negative feedback. As we'll see, seemingly impractical and even dangerous precariousness of this sort turns out to be widespread among living organisms.

"Stable" and "unstable" positions need quantitative qualifiers if one is to consider anything other than minimal perturbations. Turn a stable object far enough and it ordinarily becomes unstable, so one could plot, say, turning moment against angle, with the shift from stable to unstable marked by the angle at which the moment dropped below zero. This kind of static stability requires bearing something else in mind as well. Facile talk and thought about lateral perturbing forces can obscure the two other relevant features of a perturbation, its line of action and its moment arm.

Furthermore, for many biological situations, this static view proves inadequate. Often we have to consider dynamic stability as well. Acceleration in effect tips the direction of the restoring force, and it may shift its line of action as well. The speed of application of a perturbation commonly bears on its consequences and the effectiveness of any active response. So does the duration of a period of instability; for instance, that of any airborne phase of a terrestrial gait. (The patellar reflex, an outward jerk of the lower leg after a tap just below the knee, represents an inappropriately late response to that brief perturbation.) The activities of an organism itself can either produce or offset instability—one thinks immediately of the location of the control surfaces of swimmers and fliers relative to their centers of mass, buoyancy, and pressure. Responses to perturbation can be sluggishly overdamped or sufficiently underdamped to permit transient, sustained, or even increasing oscillations. Problems of dynamic stability have bedeviled vehicular design at least since wheelbarrows were invented, in China, over a thousand years ago.

STATIC STABILITY—SESSILE SYSTEMS

For sessile organisms well attached to substrata, the issue of stability in the present sense remains moot. Only attachment strength and vulnerability to peel failure relative to detachment moment, thus to the magnitude and line of action of lift plus drag, hold consequences for a limpet, snail, mussel, or waterpenny beetle on a rock. Only a few sessile organisms, such as flounder (plaice), manage to hold position in moving water without some secure attachment, and these latter are neither erect nor exposed to especially rapid flows. But for some of the largest sessile terrestrial organisms, remaining in place may rest on gravitational stability.

Many, probably most, terrestrial plants attach themselves to the ground with sufficient strength to resist the turning moment of wind-induced drag. But with increasing size, reliance on well-ramified, tension-resisting roots becomes ever more problematic. Greater height increases both the speed of local winds and the turning moments due to drag. And a greater area of foliage raises the drag itself. Any reasonable scaling assumption will have attachment effectiveness increasing with a lower exponent than turning moment. Some large plants do seem to manage mainly by ground-grabbing, most notably bamboos and tropical trees that can take advantage of an ample tangle of roots in the soil. Some, where I live most notably large specimens of the loblolly pine (*Pinus taeda*), limit turning with a stiff, deep, central taproot, essentially a downward extension of their trunks, as pointed out in chapter 7.

Chapter 7 also considered a tree that resists uprooting, not by attachment to the ground, but by being gravitationally stable. With ample weight, a low or deflection-resistant center of gravity, and a wide, stiff, partially buried base, the "up-" in "uprooting" takes on especially literal meaning. When such a tree does uproot, the lower portion of the trunk commonly rests a meter or two above the ground. Plate 12.1 left shows such a tree, one growing in a fairly open and unsheltered location. An instructive variation of the arrangement has repeatedly appeared in trees that live in the shallow water of swamps. Weight near the base, where it won't move laterally when wind-loaded, increases stability most effectively. But what matters is effective weight, that is, weight less buoyant force. The densities of most fresh woods lie below that of water, and even the few denser ones are not much denser. So trunk volume below the water line has little stabilizing value—and these trees have trunk enlargements just above the water, as in plate 12.1 right.

How to Stand on Legs

A tripod can be stable if its center of gravity falls in the area defined by three straight lines joining the ends of its legs. Additional supporting legs merely increase the number of lines needed to establish that area. And they only do that if flexibility of leg length or joints or else substrate compliance allows more than three to contact the ground. Naively, then, we might assume that unipeds (as are some standing birds) and bipeds (ourselves) are unstable when standing, while quadrupeds, pentapeds (kangaroos, at times), and yet more leggy creatures can stand stably, at least if they have amply stiff joints. In fact unipeds and bipeds need not be intrinsically unstable since feet can provide sufficient contact area to circumscribe the line of action of gravitational force. Still, stability does ask that they have broad, stiffly articulated feet. The instructive exception, not hard to experience, consists of standing (not walking) on stilts—virtually impossible without fairly frequent changes in contact points.

Since no animal in nature engages in bipedal stilt-standing, no standing animal need be intrinsically unstable. Yet however easily achieved, few if any standing animals take full advantage of stable postures. We humans do appear to stand directly over our feet with our weight borne by compression of our leg bones, deliberately courting if not mandating instability. But we tilt slightly forward and then offset that shift of center of gravity by muscular action, principally as tension in the large muscles of our calves and the backs of our thighs. (Hasan 2005 describes the remarkably complex system involved.) We—and most (perhaps all) erect quadrupeds—continuously sense position and adjust the output of tonic muscles. As Sir Charles Sherrington (1906) pointed out a century ago, an elaborate proprioceptive system that signals forces and lengths of muscles does the critical sensing. We pay scant attention to its operation unless we do something mildly unnatural, such as standing for a prolonged period on one leg. Toy horses, cows, and humans stand precariously; real ones are not such easy pushovers.

Standing posture varies systematically with body size in a way that's sensible for a slightly unstable system not profligate with force and work. A large mammal stands almost perfectly erect, bearing nearly all its weight on the lengthwise bones of its legs. A small mammal crouches, with leg joints flexed to one degree or another. A crude rationale goes as follows. Muscle makes up about the same fraction of the body mass of all mammals, and the contractile force a muscle produces varies with its cross-sectional area. So, all else equal, the small mammal can rely more on muscle and maintain a less bone-supported and more unstable posture relative to its body mass. If, as in figure 12.2, a joint is flexed by an angle, θ, body weight is W_b, and muscle force is F_m,

$$\frac{F_m}{W_b} \propto \tan \theta. \tag{12.1}$$

F_m scales with the square of body length, W_b scales with length cubed. So their ratio must scale inversely with length, and for creatures of similar shape and density, force relative to body weight should vary inversely with length or with $W^{-0.33}$. Thus (with constant gravitational acceleration), flexure angle should vary as

$$\tan \theta \propto M_b^{-0.33}. \tag{12.2}$$

While I've not seen a direct test of the prediction, Biewener (2003) predicts and supports a similar one, taking a somewhat different approach. He cites data that give $M^{+0.26}$ for "effective limb mechanical advantage," close to the inverse of the flexure angle used here—the difference between 0.26 and 0.33 is unlikely to be significant. He states (and I completely agree) that, by lowering the center of gravity, a flexed stance confers advantages when an animal accelerates, both for linear and angular acceleration. In short (one might say), it may impose a cost in stability— more forceful corrective motions—but it enhances maneuverability.

One predicts, then, that the height when standing of a mammal's center of gravity with respect to body mass will vary with an exponent greater than the 0.33 of isometry. That squares with the observation that small mammals are on average long and stand low, while big ones are short and stand high.

Anticipating just a bit, the small mammal's flexed stance and concomitant change in muscle location might improve muscular efficiency in locomotion. Muscle does best if shortened slowly, as measured in muscle lengths per unit time—"intrinsic rate of contraction." But the higher

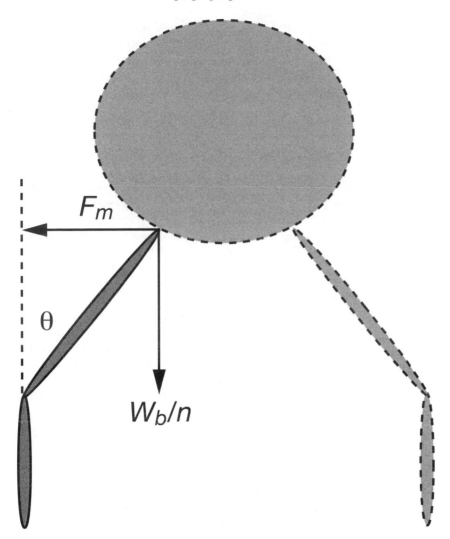

Figure 12.2. Adjoining long bones of a standing animal with the flexion angle, θ, between them. W_b is body weight; F_m is the force the relevant muscles must exert to offset the flexion; n is the number of (identically loaded) legs.

stride frequency of smaller animals normally offsets this advantage of smaller size. Add the size-dependent variation in location, and the small animal needs higher intrinsic rates and has to pay a higher price in cost of transport, energy expended relative to the product of mass moved, and distance.

Cows, as expected, stand on almost unflexed legs. A practice termed "cow-tipping" enjoys a widespread body of folklore, at least in North America. Supposedly one or a few people who sneak up at night on a sleeping, standing cow in a field can push it over, with distinctly detrimental effects on the animal; they take advantage of the narrow window of stability resulting from the high center of gravity and closely spaced legs. An analysis by Lillie and Boechler, at the University of British Columbia (2005), concludes that a standing cow has sufficient stability to require an impractically large force for such a prank, about 3000 newtons (equivalent to 300 kilograms). Thus, if a single human can push about 300 newtons at the requisite height, ten synchronized pushers would be needed. (Lillie and Boechler assume what I think is an overly generous estimate of the push a typical human can exert. Cotterell and Kamminga 1990 cite a datum for maximum pull of 280 newtons, which ought to be about right for pushing and which I've just rounded off to 300.)

Pushing force, though, may not be the key constraint that renders the stories apocryphal. More important, cows do not sleep standing up, and when standing, they have the usual dynamic instability and ever-vigilant reflexes quite noticeable if one tries to tip a dog or cat. If the cow does no more than modestly widen its stance without an overall shift of its center of gravity, about 4000 newtons or fourteen pushers would be needed— quite a challenge to deploy without angering the cow. As Young-Hui Chang tells me, even a flamingo that stands quietly on one leg above a splayed foot makes continuous minor muscular adjustments as directed by its proprioceptive system.

And How to Walk on Legs

Moving about on legs adds other destabilizing factors. Two unavoidable factors loom largest. First, progress demands pushing rearward on the substratum. So the line of action of the propulsive force will lie below that of the resistive forces of inertia (resisting acceleration) and drag (resisting speed). Second, motion most often means lifting at least one leg from the substratum, so an animal must not tip excessively when supported by one less than its normal complement. For most forms of legged locomotion requiring continuous stability would greatly limit motions and gaits, and in practice most pedalists are somewhat unstable most of the time.

The moment from the distance between an animal's rearward push and the resistance to that push turns the animal head-backwards. A forward shift of the center of gravity normally heads that off with a head-forward moment. Acceleration at the start of a walk or run takes more of a shift than does moving steadily; for most forms of terrestrial locomotion getting mass into motion greatly exceeds drag as an impediment. One leans forward with the greatest tilt when one starts walking, and walking when half submerged in water (more resistance, less gravitational force) takes extra tilt. We start walking by deliberately falling forward, and we cannot stand in quite the same posture as when walking at anything over the most modest pace. Adding resistance by asking that a person push or pull elicits an even greater forward tilt.

Increasing the number of legs from two to four does not eliminate the turning moment, although by not shifting the center of gravity to a statically unstable location it may at least simplify the situation. So asking that a quadruped push or (more commonly) pull still exacerbates instability. At one time horses were often harnessed to the carriages of the ostentatious with checkreins that kept their heads high. That both limited their pulling abilities and obviously distressed them. A nineteenth-century children's classic, *Black Beauty* (Sewell 1877) made much of that and other abusive practices and probably contributed to its abandonment.

When walking, bipeds like us sway slightly from side to side as support shifts back and forth from one leg to the other, again displacing ground contact from beneath the center of gravity. Again, one can't stand in most positions one assumes while walking at all but the slowest speed—snapshots of standing postures can't be arranged into a walking sequence. Thus we deal with lateral as well as fore-and-aft instability, although the lateral kind averages out over time. At least lateral tilting should shift the line of action of the center of gravity closer to the ground-contact point of the next leg that will prop and push. Penguins, relative to their heights the shortest-legged of birds, sway the most. For them the gravitational shift becomes central to a pendulum-like cost-minimization scheme (Griffin and Kram 2000). In quadrupeds, analogous sway characterizes a gait called "pacing" or "racking" in which both right and then both left legs move simultaneously. Horses can be taught to pace, while long-legged giraffes and camels normally pace, probably because it lets these long-legged animals swing their legs farther without front-hind interference. Pacing, of course, imposes the same instability as bipedal walking and so throws away a major advantage of quadrupedality.

A variable called "duty factor," the fraction of the time a given leg provides support, helps us judge whether locomotion can be stable. A two-legged walker cannot be fully stable even at a duty factor of 1.0—standing still—as already noted. Oddly enough, that doesn't demand that bipedal walkers have a continuously vigilant and actively intervening proprioceptive system. Small wind-up walking toys or similar unpowered downhill walkers do quite well, although usually with feet of biologically

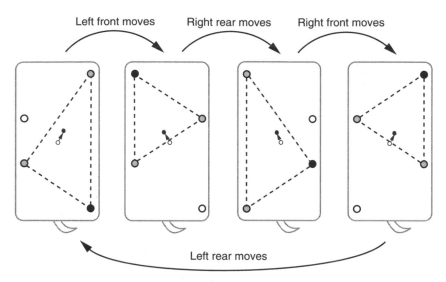

Figure 12.3. The normal stepping pattern for a quadrupedal walker. The most recent footfall is shown dark, the one just lifted is light. Arrows indicate shifts of the center of gravity needed to move it from the center of a standing stance to equivalent positions with respect to a line between diagonally opposite supporting legs.

unreasonable areas. Their centers of gravity may never lie directly above a leg, but properly timed foot-falls limit tilting to either side. Coleman and Ruina (1998) have devised a more elaborate version of such a self-compensating downhill walker, a physical model that can be put together from widely available toys ("Tinkertoys"), and provided a theoretical treatment of such stability-while-in-motion. Powered versions (Collins et al. 2005) with only the most minimal control walk with efficiencies comparable to those of walking humans. The particulars can be viewed at www.news.cornell.edu/releases/April98/tinkertoy.walker.ssl.html and ruina.tam.cornell.edu/research/topics/locomotion_and_robotics/papers/tinkertoy_walker/index.htm.

A four-legged creature can be stable, but only if it never lifts more than one leg off the ground, that is, if the duty factor equals or exceeds 0.75. And it must shift its center of gravity slightly, typically by tilting the body away from whichever leg is held aloft, as in figure 12.3. That keeps the center of gravity over the triangle formed by the contact points of the remaining legs. In effect, a leg must give the ground a slight body-tilting push just before breaking contact. Quadrupeds do walk stably (postural reflexes aside), but mainly when at their slowest, as when stalking prey. We might expect slow walking to demand rather than just permit static stability, but work on chelonians—turtles and tortoises—says quite the opposite. While duty factors generally run well over 0.75, Jayes and Alexander (1980) found that at times only two feet contacted the ground. They gave a persuasive (if counterintuitive) argument that eschewing stability permits slower and more efficient muscle action.

But slower locomotion does lower tolerance of instability. As Alexander (2003) notes, while forces need only balance when averaged over a stride, during any unstable period an animal falls with gravitational acceleration, g. And some relatively constant fraction of leg length, h, must limit permissible falling distance. As an indicator of the need to preserve stability, he suggests a dimensionless expression based on these variables and on stride frequency, f:

$$\frac{g}{2f^2h} \tag{12.3}$$

Thus instability becomes more tolerable as an animal increases its stride frequency, especially if done without much concomitant decrease in leg length. Put in practical terms, at low speeds and thus low stride frequencies, stable gaits work better; as speeds increase, unstable gaits become ever more practical and higher degrees of instability more tolerable. Dogs can tolerate a lot of instability when galloping but less when walking; turtles, low to the ground and making infrequent strides, should be much less tolerant of instability—although, as just described, they can still be unstable.

Incidentally, Alexander's (2003) approach parallels a suggestion I made (Vogel 2003) about the minimum speed for galloping, also based on maximum fraction of leg length that an animal can fall between footfalls. I invoked the Froude number as a predictive variable; equation. (12.3) amounts to the reciprocal of the Froude number (Fr) if animals swing their legs similarly, so speed is proportional to the product of stride frequency and leg length, fh:

$$Fr = \frac{v^2}{gh} \tag{12.4}$$

Six legs permit unconditional static stability. A hexapedal animal need only support itself on alternative triangles, and a duty factor of 0.5 is ample. Insects, paradigmatically hexapedal, use such a stable gait at low speeds but become increasingly unstable as they move faster (Ting et al. 1994). (Of course not all insects walk on all of their six legs—for instance, praying mantises and mosquitoes use only four.) At the highest speeds some insects, such as cockroaches and ants, have fully aerial phases (Full et al. 2002)—as we do when running. Further increase in number of legs further reduces the minimum duty cycle consistent with stability, but six are the fewest paired legs with the option of fully stable walking without center-of-gravity shifting. That has stimulated considerable interest in how insects walk and run by designers of walking robots and robotic vehicles.

THE STABILITY OF AIRCRAFT, LIVING AND NONLIVING

An object standing or moving on the earth's solid surface faces two planes of potential instability, both of them vertical, from having its center of gravity above its contact with the substratum. For bilaterally symmetrical movers these planes are side-to-side and fore-and-aft. Alternatively, we can adopt anatomical practice and designate the planes transverse and sagittal.

Devices moving through continuous media in addition face a horizontal plane of potential instability, frontal to the anatomist. For names we commonly adopt ones from aviation, perhaps to emphasize that we mean (as that community does) changes in orientation (rotation) rather than position (translation). As in figure 12.4, that community calls side-to-side or transverse vertical tipping "roll," fore-and-aft vertical rotation "pitch," and side-to-side horizontal swinging "yaw."

Dealing with three planes complicates both achieving directional control and analyzing how organisms might be managing—as it amplifies the issue's importance. (In fact, perturbation about one axis commonly affects orientation about a different axis, introducing another element of instability.) Putting aside degree of stability, after a perturbation a stable aircraft returns to its previous orientation without active adjustment of its controls. An unstable one either does not return or deviates even further.

In a short article in *Evolution*, in 1952, John Maynard Smith drew attention to the substantial orientational instability of most extant flying animals. Before taking a degree with the great evolutionary biologist J.B.S. Haldane, Maynard Smith had spent some time as an aeronautical engineer, so he brought a new perspective on flight to both paleontologists and physiologists. He compared present-day insects and birds as well as mature pterosaurs to the earliest then-known flying ancestors of each. (About early bats, information was inadequate.) He suggested that, as active flight developed in each lineage, evolution produced ever less stable designs. In particular and as with aircraft, control in pitch presented the greatest challenge. Most early forms had greater development of horizontal surfaces on their rear portions—lateral plates on insects such as the dragonfly-like Paleodictyoptera, lateral membranes on the pterosaurs, and large tails on the birds—that stabilized flight much as rear

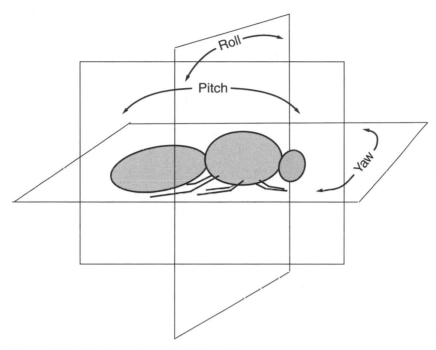

Figure 12.4. The three planes in which a flier can turn.

fletching stabilizes arrows. He pointed out that, while many extant birds have large tails, most mainly deploy them just for take-offs and landings, when tails lower the stalling speed and thus facilitate operation without runways. (The remarkable pictures in Dalton 2001 certainly bear that out.)

What permitted instability to increase was the concurrent evolution of ever more competent flight control, that is, of neuromuscular systems. What drove the process each time was the inherent trade-off in flying machines between stability and maneuverability—the same situation faced by legged locomotion on land in a much less forgiving guise. Contemporary birds are quite unstable, rendering them dangerously flawed models for human aircraft. Many early attempts to build airplanes foundered on inadequate appreciation of that awkward reality (Harris 1989). At least one case proved fatal, that of Otto Lilienthal, in 1896, author of the pioneering analysis, *Bird Flight as the Basis of Aviation* (1889), and pilot of what we would call hang-gliders. Others, notably Samuel Langley, took great pains to assure pitching stability—the most troublesome plane and the one that doomed Lilienthal. Langley's best full-scale airplane, which failed (at least) from structural weakness, achieved inherent pitch stability with tandem wings, one pair behind the other.

Unlike Lilienthal and others, the Wrights gave considerable attention both to stability and to control by adjustments of the aircraft rather than the position of the aviator (Culick and Jex 1987). Their later gliders and 1903 Flyer had sufficient stability to be safe and reliable in breezes but not so much as to compromise control. Indeed, the only fundamental change made since consists of substituting for the canard wing in front the now familiar pitch-controlling horizontal tail. Canard wings persist only in a few high-performance aircraft (usually with fast-acting computer control offsetting intrinsic instability) and possibly as the "hammerheads" on sharks in the genus *Sphyrna*. On the latter, they may (a good study is overdue) facilitate following the contour of an irregular sea bottom or lunging suddenly downward. A front paddle directed obliquely forward by white-water canoeists approaching rocks, the "bow-rudder," works the same way except in yaw.

Stable fliers occur in nature, as one might guess, where active control is unavailable. As with terrestrial stability, plants provide the exemplars. A descending, autogyrating samara of a maple or other tree must be able to recover from the perturbations of wind gusts or branches encountered on route. Non-autogyrating gliders, closer to airplanes or gliding birds, are rarer than samaras. The most famous is the gliding seed-leaf of the Javanese cucumber *Alsomitra* (formerly *Zanonia*). As with so many phylogenetically odd animal gliders it lives only in the understory of the especially high dipterocarp forests of southeast Asia—perhaps because its nearly still understory air extends over an unusually great height range. It provided a model for the aircraft of the Etrichs, who built a series of *Zanonia*-winged craft, beginning with gliders (figure 12.5), in the first decade of the twentieth century. Most likely reacting to Lilienthal's death, they wanted assurance of stability, which they achieved with a glider that proved almost unmaneuverable (Bishop 1961). Vincenti (1990) provides a fine historical perspective on how appreciation of the issues involved gradually developed.

For stability, *Alsomitra* sacrifices straight-line performance. The so-called lift-to-drag ratio of an airfoil, hydrofoil, or propeller blade helps us see that trade-off in context. In effect, such a device generates lift (L), a force normal to its motion through the medium, at the price of an increase in drag (D), the force tending to retard its motion. The lift-to-drag (originally the more euphonious "lift-to-drift") ratio not only represents a kind of efficiency, but it translates directly into the range a glider can go in still air. It does this by setting the "glide angle," θ, the angle with respect to the horizontal, at which a passive craft will descend:

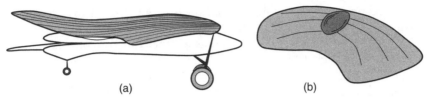

Figure 12.5. (a) The Etrichs' glider of 1906, traced from a photograph of a full-scale model at the Owl's Head Transportation Museum, Rockland, Maine; struts and cables have been omitted. (b) The seed-leaf of *Alsomitra*.

$$\cot\theta = \frac{L}{D} \quad \text{or} \quad \tan\theta = \frac{D}{L}. \tag{12.5}$$

Gliding thus maximizes distance by maximizing L/D, which varies only a little with speed.

The *Alsomitra* seed-leaf in nature develops a relatively poor lift-to-drag ratio, 3.7. In a wind tunnel, its best (with some loss of stability) is still a mediocre 4.6, about half of what an insect wing under equivalent circumstances can reach. Furthermore, its swept-back wings put the seed and center of mass ahead of the aerodynamic center (the point of action of the lift-drag resultant vector), so it has an equivalent of a stabilizing tail. As a result, the increased lift of any inadvertent upward pitch of the nose is far enough aft to offset rather than amplify the perturbation. And normal operation at an angle of attack well below the stall point provides an extra margin for that stabilizing effect (Azuma and Okuno 1987; Azuma 2006).

By contrast, birds fly as remarkably unstable craft—one cannot easily launch a stuffed bird on a smooth and fairly flat path. As mentioned, their use as models probably hampered the development of airplanes. Still, the degree of instability varies a lot; and, again, varies more or less with maneuverability. The *Alsomitra* seed-leaf accepted a lower-than-ideal lift-to-drag ratio to gain stability; birds make analogous departures from best performance to gain maneuverability.

Other things equal, a longer, narrower wing gives a higher ratio than does a shorter, broader wing—again something glimpsed but not always appreciated by earlier aircraft designers. Why, then, do sea birds have particularly long, narrow wings with pointed tips while those that glide over land have shorter, wider ones with splayed primary feathers? These terrestrial gliders apparently accept a lower best L/D—in effect a lower efficiency—and thus steeper minimum glide angles so they can turn more sharply and fly in tighter circles. Turning radius should be important when gliding in an ascending thermal torus or an updraft over irregular terrain. Stabilizing against yawing—more important in the more erratic winds over land—has been suggested as another function of splayed primaries (Pennycuick 1975), but with little hard evidence of either the action or its utility.

Splayed primary feathers help terrestrial birds fly slowly without stalling. With them, they can tolerate without stalling greater angles between wing and oncoming air, which boosts lift, which varies with the square of speed but must always equal body weight. That improves their ability to land and take off at near zero airspeed. (Norberg 1990 give a good review of the contrast.) Further enhancing that low speed performance, terrestrial soarers have lower wing loading (W), the ratio of weight (mg), to wing area (S):

$$W = \frac{mg}{S} \approx \frac{L_{req}}{S} \propto \frac{v^2}{S} \tag{12.6}$$

Weight to be supported translates into lift—roughly, since lift is strictly defined as normal to flight direction rather than to the horizon. So lower wing loading implies a lower weight-sustaining minimum flight speed.

Since in equation (12.6) m scales with l^3 and S with l^2, L_{req}/S scales with linear dimensions—lift here is lift required to support weight, not lift as proportional to wing surface at constant speed. That scaling makes wing loading higher for a larger but otherwise similar flier. Such scaling underlies not only the nonsimilarity of fliers of different sizes but also the higher flying speeds of larger craft—with their associated reduction of maneuverability, their higher takeoff and landing airspeeds, and their higher weight-specific power requirements for sustained flapping flight. Thus the 70-kilogram extinct bird *Argentavis magnificens* had to soar rather than just flap and could live only in a region of steep slopes and high winds (Chatterjee et al. 2007).

While equipped with nothing analogous to the splayed primaries of terrestrial soaring birds, the same trade-off has been recognized in bats by Aldridge (1986), Norberg and Rayner (1987), and Dietz et al. (2006). Bats that fly through forests have shorter, broader wings, and they weigh less relative to wing area so they can fly more slowly—paying a price in power (in effect overall performance) for that maneuverability. By contrast, bats that fly in open areas have longer, narrower wings, more weight relative to wing area, and they fly faster and more economically.

Like birds and bats, flying insects are fairly unstable. Extant forms lack aerodynamic stabilizers such as tails or abdominal protrusions; in any case both would be of limited use during hovering. Hind legs sticking out into the airstreams of the wings seem to provide some ruddering in some forms, and mobile abdomens (as in many wasps) at least raise the possibility of adjusting centers of mass and drag. As Dudley (2000) notes, the way the wing stroke centers above the body, with wings almost (or actually) touching at the top but not the bottom of the stroke, provides some degree of stability in both roll and yaw. (Any torque induced by a center of lift and thrust above the center of drag should matter little for a flier that can control the direction of the resultant of lift and thrust.) And the elongation of bodies fore and aft—heads and abdomens, the latter sometimes quite long—should give a bit of pitch stabilization, at least for transient perturbations.

The issue of stability may bear both on the origin of flight and on the subsequent evolution of flying lineages. As pointed out by Ellington (1991) (and summarized by Dudley 2000), a long, circular cylinder, held obliquely, can descend at glide angles as gentle as 40°, no worse than some gliding but nonflying vertebrates. Instability, though, wrecks the scheme at all but Reynolds numbers (length times speed, in effect) still lower than those of flying insects. An aerodynamic center in front of the center of gravity causes the difficulty. Any upward pitch will be magnified until, at equilibrium, the cylinder will descend vertically while oriented horizontally, parachuting rather than gliding. Small winglets, minor cuticular extensions, protruding from the sides and located toward the rear fix the problem, stabilizing the cylinder in pitch. (Moving the center of gravity forward, where it is in extant insects, can help also.) Yawing stability can be achieved by adding a caudal filament, and roll stability takes only a bilateral pair of diagonally rearward-pointing filaments instead of a single one. Of course overdoing the rear appliances ends up producing the equivalent of a fletched arrow, which both descends and orients vertically, dropping with no horizontal range and maximum speed.

In effect, a flier picks some combination of three variables—stability, maneuverabilility, and performance—each in practice multidimensional. As we've seen, both the exceptional stability of the *Alsomitra* seed-leaf and the maneuverability of terrestrial soarers come at a price in level-flight performance. In general, an increase in any one of the three variables extracts a price with respect to one or both of the others. None, though, lend themselves to definitions that combine precision with practicality, and designers face no definitional limit to what a particular combination can do. As David Alexander (2002) reminds us, birds and bats may be less constrained in that choice than insects and human aircraft since they can quickly and extensively vary wing geometry.

The Inputs for Aircraft

Only with adequate control systems can instability be tolerated, much less capitalized upon. Proprioceptive feedback loops, gravity and acceleration detectors, and associated anticipatory and dynamic devices permit walking, running, climbing, and all manner of terrestrial acrobatics. But however complex these tasks, they pale before those of control in continuous media, especially in air, far less forgiving than water. Again, the less stable the flier or swimmer, the greater its dependence on control. Moreover, even a stable flier faces a problem practically unknown to those of us with feet on the ground. Air and water often move with respect to their solid substrata. Thus, sensing what goes on in an animal's immediate vicinity may limit clues to its motion relative to the earth. And the slower the animal, the lower will be its speed relative to motion of the medium itself. So the problem bedevils organisms more than it does our boats and aircraft. Both ground speed and direction can be enigmatic, with heading giving only a limited cue about course. Face north and move forward at 1 meter per second in a wind or current from the east at the same modest speed, and you progress northwest with a ground speed of 1.4 meters per second.

Small airplanes (excluding high-performance military craft) are, by design, about as stable as possible without overly compromising their ability to obey instructions to change direction. The new pilot learns, in the words of Molly Bernheim (1959), that when things go awry, "Let go! The airplane can look after itself better, now, than you can do! Turn it loose! Then, and only then, you may guide it gently where you want it to go." Even for airplanes, airspeed and ground speed may be quite different, and heading may not equal course. Moreover, our land-based sensory equipment can mislead us. For instance, semicircular canals can't reliably separate gravitational from angular accelerations. So a banked turn feels no different from straight flight, and the pilot must read the instruments rather than the receptors medial to the seat of the pants.

How do animals get the sensory input critical for active control? Visual signals provide widely used references for both orientation and location, and both birds and insects typically devote large areas of brain to processing visual input. Horizontal cells have been known in bird retinas at least since the work of the great neuroanatomist, Ramón y Cajal, a century ago. They purportedly select horizontal lines for attention, horizon-detection that distinguishes level flight from banked turns. Were night-flying birds, insects, and, of course, bats not so common and accomplished, one might declare visual input essential—which it certainly cannot be. At least equipping an enclosed volume of air to work as an altimeter should be simple enough if a flier needs comparisons that only span brief periods. A human acquires chambers that can be painfully effective for those taking airplane flights with plugged Eustachian tubes. And, analogously, swim bladders can signal depth changes in fishes.

The flight motors of insects, on which work has been extensive, have at least three additional sources of sensory input. Bending of antennae and setae equipped with mechanoreceptors at their bases provides information about local airflow, probably including the airflow on each side caused by the wings themselves (Humphrey and Barth 2008). Additional mechanoreceptors on the wings and in cuticular structures adjacent to them can signal what the wings are doing. In addition, because all the relevant structures are nonrigid, they can provide feedback on the loads the wings encounter. Since oscillating wings act as gyrosensitive devices, the receptors should receive usefully dynamic inputs. Several groups of insects, most notably the true flies (Diptera), have developed gyrosensitivity further, converting one pair of wings (in Diptera the hindwings) into stalks with knobs on their ends that still oscillate like wings. Finally, the flexible connections between thoraces, housing the flight motors, and both heads and abdomens, permit both the latter to provide inertial information—for instance, during active or passive turning. Taylor and Krapp (2008) give a good sense of the present state of the art.

The principal difficulty for our understanding of how flying animals manage and probably for the animals themselves is the lack of an obvious source of earth-referenced data. Vision cannot form the sole such source, but sensing, for instance, cuticular deformation can do little to augment it. The need has to be acute in a domain in which ambient winds rival or even exceed flight speeds and in which the variations that we lump as local turbulence can't be easily averaged over time or space. People investigating bird migration—which I want to skip over here—have wrestled with the problem for many years. The same problem for both animal and biologist afflicts fishes that hold position in murky, moving water (Howland and Howland 1962). The question boils down to whether (and of what accuracy and reliability) animals have on-board GPS ("global positioning system") gear—as even small airplanes now rely upon.

SWIMMING

Submerged swimming, the usual animal mode, suffers the same general trade-offs and affords the same opportunities as flying. Still, a more forgiving locomotory mode, it permits a wider range of designs and solutions. For a swimmer of a given size to swim a given distance costs less than for a flyer to fly or a walker to walk. Per unit time—since hovering costs next to nothing—it costs vastly less than any form of active flying. Only soaring, essentially gravity-sustained, comes close. However economical, though, both predators and prey do the same cost-benefit calculation. And suspension feeders that actively swim (or pump) face the outcome of success by other present and previous suspension feeders. So once again systems must to some extent pay for maneuverability in the currency of performance efficiency.

Most swimming vertebrates have a particular non-locomotory instability with little parallel among fliers. Achieving buoyancy with a non-rigid, gas-filled container makes them unstable with respect to depth. Increased depth compresses the gas, reduces buoyancy, and impels an animal to go still deeper; likewise, decreased depth urges further decrease. The problem and its various solutions formed a large part of chapter 8.

That gas bladder generates a more subtle problem. If the center of the water displaced by a submerged object lies below its center of gravity, then the object prefers to be inverted. With the usual convention, we declare the vertical distance between those centers, the so-called metacentric height, negative. Although the bladder has moved from its ancestral ventral location (like lungs) to a position above the gut, most bony fish with swimbladders still have negative metacentric heights. So their normal postures are normally unstable—a dead fish ordinarily goes belly up. Fish that live in moving streams seem to be the most unstable, perhaps because they have to expend power to hold position anyway. Fish that live in still water and do not swim continuously tend to be less unstable. "Resting" may be easier than swimming, but a stationary fish can't trim a hydrodynamic surface to adjust its position and thus lacks a mode of active stabilization that might offset this static instability (Webb 2002).

Both fish and cetaceans add more instability by moving through water. As we saw in flying animals, stability and maneuverability are again to a considerable extent antithetical. The trade-off of speed for maneuverability appears even clearer for the swimmers than for flyers. In particular, stiff-bodied forms tend to be both faster and less maneuverable than flexible ones, whether one compares cetaceans (Fish 2002) or fishes (Webb 2002).

Swimmers most often propel themselves with drivers at their downstream ends, whether the fins of fishes, the flukes of cetaceans, or the jets of cephalopods. (But not always—some fish use pectoral fins or opercular jets while penguins, some other birds, and sea lions, to mention a few, use modified forelimbs.) Pushing from behind rather than pulling from in front generates what we might call dynamic instability. Any inadvertent yaw makes the propulsor generate a turning moment that amplifies that initial yaw. Still, this doesn't seem to raise any noteworthy difficulty despite the inherent instability of pushes that (except in jet-propelled forms) alternate from one side of the body to the other. Nor despite the analogous instability and consequent bad behavior of rear-propelled, rear-heavy automobiles.

Another form of instability comes from trimming controls in front of centers of gravity. Hammerhead sharks, as noted earlier, may use the hammerhead as a canard wing, a severely destabilizing device, to improve their ability to swim at a fixed distance above a nonlevel substratum. Skates and rays may do the same with their relatively anterior "wings"—many do swim just above the substratum. The subterminal rather than (as in bony fishes) terminal mouths of all these elasmobranchs suggest a bottom tracking, swimming ancestor. Less extreme are ordinary pectoral fins and flippers, but in both fishes and cetaceans even the positions of these are usually ahead of the centers of both buoyancy and gravity (Fish 2002).

In both fliers and swimmers, large forms that chase small forms for food tend to be slower, less stable, and more maneuverable than closely related large forms with other modes of feeding. Large predators retain a speed advantage but must offset worse minimum turning radii and maximum turning rates. Interspecific comparisons among toothed whales point up that trade-off (Woodward et al. 2006). Humpback whales, notably acrobatic baleen whales, have unusually large pectoral flippers and are not especially rapid swimmers (Nowak 1991) by large-whale standards. The flight motor of dragonflies, large aerial predators, drives the wings directly rather than indirectly, as in most other insects, giving them an unusual degree of independent control of their four wings (D. E. Alexander 1986) but with efficiency-reducing interaction.

Surface swimming runs into the problem of negative metacentric heights in a particularly nasty form. Floating high in the water virtually guarantees a highly negative value unless the craft has a lot of ballast deep in the hull. At least the problem can be ameliorated with a scheme unavailable to submerged swimmers. Most ships have V-shaped or U-shaped hulls. This geometry requires that the center of gravity of the craft be lifted if it rolls either way from exactly upright, supplying a restoring torque. Counterintuitively, perhaps, a broad, flat bottom with sides that then slope inward gives a craft that lacks that region of stability. The few animals that swim on the surface (with displacement hulls, not planing or using surface tension) typically have hulls that taper downward like our boats—something especially conspicuous in young sea turtles (Wyneken, personal communication).

THE 'FLIGHTS' OF NONFLYERS

In the end, all fliers achieve stability with aerodynamic devices—adjustable wings, deployable tails, and so forth. Once a flier has propulsive or gliding appendages and proper sensors, control needs only modest equipment. But what can a nonflier do if it finds itself in midair and prefers to land in a specific orientation? While hydrodynamic control can be effected by fairly small structures, only tiny nonfliers can get effective aerodynamic service from ordinary appendages. Some ants that live high in tropical forests make stable flying jumps at respectable glide angles (moving abdomen first, incidentally) by orienting their legs for aerodynamic stabilization (Yanoviak et al. 2005). Larger nonfliers must play with angular acceleration and velocity in a world that awkwardly insists that angular momentum be conserved.

Just as the product of mass and linear velocity gives ordinary momentum, the product of moment of inertia, I, and angular velocity, ω, equals angular momentum, H:

$$H = I\omega. \tag{12.7}$$

Moment of inertia, the second moment of mass, is the sum of the elements of mass, m, times, for each element, the square of its distance, r, from the axis of rotation:

$$I = \Sigma mr^2 = \int r^2 dm. \tag{12.8}$$

Constancy of angular momentum means that a change in angular velocity comes with an inverse change in moment of inertia (eq. 12.7). In a world that also insists that mass be conserved, changing moment of inertia depends on changing the effective overall r, the "radius of gyration" (eq. 12.8).

Increasing angular velocity by decreasing radius of gyration—a figure skater or ballet dancer does that by drawing arms and legs closer to torso and thus to the axis of rotation. Alternatively, an external contact can yield a moment that imparts angular velocity. A springboard diver can do that in at least two ways (Frolich 1979). A run out along the board gives the whole body a translatory velocity, but the jump at the end, besides imparting an upward component to velocity, slows the lower part of the body (figure 12.6a). So the diver takes off with some angular velocity and angular momentum. Tucking in legs and arms in midair then increases angular velocity. With (as conservation requires) no change in angular momentum, a somersault ensues. Alternatively (or in combination), the diver may jump with the body tilted forward so the resulting torque of the vertical push imparts an initial angular velocity (figure 12.6b). Either way, the diver reduces angular velocity again before entering the water by extending arms and legs.

Does conservation of angular momentum require that any mid-air rotation trace back to an initial, visually subtle, angular velocity? If by

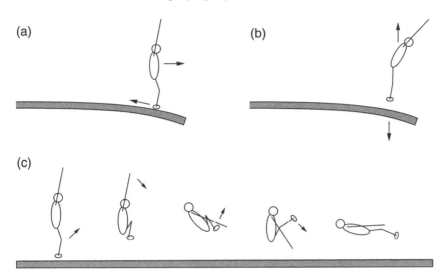

Figure 12.6. (a) Turning during a springboard dive by making the feet lag the torso. (b) Turning by taking off with head and torso forward of the final contact point. (c) The sequence in which appendages are moved to perform a zero-angular-momentum back quarter somersault, as described by Frolich (1979).

"rotation" one implies time-averaged net angular momentum, then (without aerodynamic or Coriolis effects) that must be correct. What's less self-evident is that changes in body orientation—rotation about any axis—are possible without net change in angular momentum of the body as a whole. Such orientational changes have been unequivocally demonstrated in springboard divers, acrobats, trampoline jumpers, space walkers, and falling cats (as well as in some other mammals). Frolich (1979) and Edwards (1986) provide good descriptions and analyses of how it can be done, Brancazio (1984) gives a quick summary, and Stewart (2004) shows a nice set of color photographs of a falling cat.

Figure 12.6c, adapted from Frolich (1979), shows a diver or gymnast reorienting in a pitching plane about the body's long axis. Initially the body extends full length. Tucking legs up close to the torso reduces the moment of inertia—not in itself inducing much orientational change, but amplifying the subsequent change. Swinging the arms forward and down against the body and thus giving them angular velocity and momentum gives the rest of the body an equal and opposite angular momentum and an opposite, if lesser, angular velocity. When the arm motion stops, the motion of the rest of the body has to stop. But the body has shifted orientation substantially—up to about 80°. The arms may then be reextended by moving them along the body axis or swinging them outward to restore the original extended posture. Since moment of inertia is a second rather than a first moment—incorporating r^2 rather than r—means that even fairly light appendages can be useful, as long as they're reasonably long.

Note a characteristic—and diagnostic—difference between turns that take advantage of initial angular momentum and those that manage despite zero angular momentum. In the first, both angular momentum and angular velocity remain, so the body keeps rotating unless stopped by some external agency. In the second, with no overall angular momentum, the body stops turning when some parts stop moving relative to others. One can experience both in a swivel chair. Initial angular momentum just takes a push against floor or desk, what one ordinarily does in such a chair. But one can turn, say, counterclockwise, by extending the arms, swinging both clockwise, drawing them back against the torso while moving them counterclockwise, extending them again, and repeating—each time progressing a few degrees. Holding weights in the hands increases the effectiveness of the maneuver. (As an exercise, the reader might attempt to explain how a child can put a swing in motion.)

Domestic cats, famously able to land on their feet, do just such zero angular momentum turns as they fall. Cats can reportedly turn 180° around their long axes during a 1-meter fall, which takes less than half a second. Peak head acceleration (where the turning begins) may exceed 120,000 degrees per second squared (O'Leary and Ravasio 1994). Tailless toms trail tailed toms in tests. According to Kane and Scher (1969) and Edwards (1986), and as in figure 12.7, the supine cat begins by arching its back so the whole animal is concave upward. It then twists the body about the vertebral column, beginning with the head, while maintaining that downward concavity, until the whole torso faces downward. Finally, it straightens the back again, halting rotation.

Dogs, less limber, are less adept at righting; in one informal test, a dachshund failed completely and took umbrage at the imposition. Rats and many other small mammals, though, right themselves quite competently. The behavior not only ensures landing on properly shock-mounted appendages to lower deceleration, but it also must increase drag during long descents. That will reduce both terminal velocity and the rate of approach to terminal velocity.

At least in a class with cats are geckos. While spectacularly rapid climbers and famous adherers, these little lizards do lose purchase from time to time. If that happens, they put their tails in motion and perform zero angular momentum turns in as little as a tenth of a second, thereafter falling (or parachuting) to controlled, foot-first, landings. Gecko tails play more critical roles than cat tails—tailless animals (they commonly appease predators with a taste of tail) can't right themselves at all. All

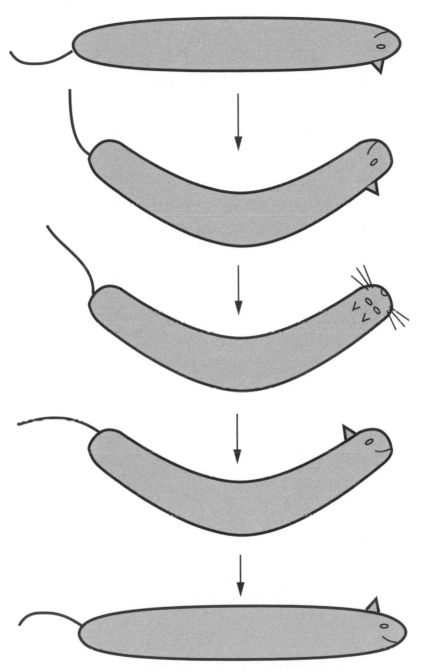

Figure 12.7. Righting of a mammal's torso with a zero-angular-momentum twist, as explained by Edwards (1986). In practice the twist begins at the head, the tail counterrotates, and movement of the legs (in a manner analogous to that shown in figure 12.6c) plays at least a supporting role.

this from a recent paper by Jusufi et al. (2008)—which has lovely pictures and references to movies.

Spinner dolphins (*Stenella longirostris*) make spectacular upward leaps, rotating while airborne as many as seven times about the long axes of their bodies. But, by contrast with cats and geckos, they make no significant use of zero angular momentum turning; instead, they drive their aerial turning by asymmetrical motion of their flukes just prior to emersion. Remoras seem to be dislodged during reentry, perhaps the function of the maneuver (Fish et al. 2006). One might expect that the high drag an animal experiences in water makes underwater inertial turning at once ineffective and unnecessary. But we should not dismiss the possibility out of hand. Photographs of the so-called pinwheeling maneuver of bottlenose dolphins (*Tursiops truncatus*) (Maresh et al. 2004) look to my eyes strikingly similar to photographs and diagrams of righting cats.

What about other cases of righting with zero angular momentum turns? An unfledged bird falling from a nest probably has sufficient plumage to keep it from reach a hazardous terminal velocity—the speed at which drag, speed-dependent, reaches weight. And most nonflying, non-mammalian gliders such as flying frogs and flying lizards can exert aerodynamic control. An exception might be flying snakes (genus *Chrysopelea*), which while gliding downward do quite a lot of mid-air writhing and maneuvering (Socha et al. 2005).

The relative utility of inertial and aerodynamic mid-air turning depends on both airspeed and body size. Faster motion favors reliance on aerodynamics, with both lift and drag increasing with something close to the square of speed. Larger size favors inertial turning due to the concomitant reduction in surface-to-volume ratio. Humans can, as we have seen, do quite well at inertial turning. But aerodynamic effects should not be casually dismissed since large animals fall more rapidly than small ones (if less so than one expects, as explained in chapter 6). Moreover, significant use of inertial turning has recently been shown (along with aerodynamic turning) in flying birds (Hedrich et al. 2007). Most birds perform downstrokes with extended wings and upstrokes with somewhat flexed wings. Flying straight generates no overall difference in moment of inertia since the wings cancel each other's asymmetry. But when turning, the outer wing increases amplitude, which will roll that side upward, aiding the extra aerodynamic lift but without extracting a fully compensatory price in drag.

One odd convergence in small mammals suggests routine use of aerodynamic turning. In at least five lineages, long-hind-legged jumping animals have long tails (longer than head+body) with tufts of hair on their ends. Three of these are rodents—kangaroo rats (*Dipodomys*; Heteromyidae), some gerbils and jirds (*Gerbillurus, Meriones*; Muridae); and jerboas (*Dipus, Jaculus*, etc.; Dipodidae). Two are marsupials—the kultarr (*Antechinomys*) and an extinct South American form (Simpson 1970; Mares 2002). Comparably long-tailed animals without long hind legs typically lack such terminal tufts, judging from the photographs in Nowak (1991). Movies of kangaroo rats engaged in intraspecific interactions show extended tails flung vigorously in all directions, with the tufts clearly visible. (See, for instance, Disney 1953.) Still photographs taken under comparable circumstances often show erected tail hairs (Schmidt-Nielsen, personal observation). Perhaps the tuft, especially when erected, increases the drag of a tail moved laterally, and drag so far from the body provides torque that helps turn the body in the other direction. Thus aerodynamic and inertial devices might combine in aiding mid-air maneuvering—but I'm not aware that the phenomenon has been investigated.

List of Symbols

SI units are assumed throughout the text. Note that certain symbols serve several variables—I've tried to follow common practice, and the chapters draw from several ordinarily distinct traditions; in context little ambiguity should result.

Ba	Bagnold number, Baudoin number
Bo	Bond number
C_d	drag coefficient
C_l	lift coefficient
C_p	heat capacity (specific heat) at constant pressure
C_p, C_{pt}	pressure coefficient, conventional
C_{pl}	pressure coefficient, assuming laminar flow
C_s	molality of solution
c	speed of light
c_p	specific heat (energy per mass times temperature change)
D	drag
D	molecular diffusion coefficient
d	distance
E	Young's modulus of elasticity
e	efficiency
F	force
F_c	Coriolis force
F_g	Gravitational force
Fr	Froude number
f	frequency
G	shear modulus of elasticity
Ga	Galileo number
GHI	Gravitational hazard index
Gr	Grashof number
g	gravitational acceleration
H	angular momentum

h height above the ground

I moment of inertia (rotating body)

I second moment of area (structure)

J advance ratio, propeller

J second polar moment of area, loaded structure

K bulk modulus (1/compressibility)

K the Boltzmann constant

K_s salting constant

k thermal conductivity

L lift

l length

M_b body mass

m mass

N Avogadro's number

n number of any item

Pe Péclet number

p pressure

Q total flow (volume per time)

Q_{10} reaction rate increase factor for a 10-degree temperature increase

q rate of energy transfer

R gas-law constant

Re Reynolds number

r radius, radius of gyration

S cross-sectional (or other) area

S solubility

Sc Schmidt number

T temperature, Kelvin (default) or Celsius

t time

V volume

v velocity (or speed)

W weight

x distance (in the x-direction)

y distance (in the y-direction)

β volumetric thermal expansion coefficient

γ surface tension

Δ change in value of a variable

δ velocity gradient ("boundary layer") thickness

ε radiant emissivity

ε glide angle (descent, relative to horizontal)

θ angle

ϑ latitude

λ wavelength

μ viscosity

ν Poisson's ratio

π 3.1416 (circumference/diameter, circle)

ρ density

σ stress (force per area)

Ω angular velocity, earth

ω angular velocity

References and Index of Citations

Aerts P 1998 Vertical jumping in *Galago senegalensis*: the quest for an obligate mechanical power amplifier; *Phil. Trans. R. Soc. Lond.* **B353** 1607–1620 (*48*)

Agnisola C 1990 Functional morphology of the coronary supply of the systemic heart of *Octopus vulgaris*; *Physiol. Zool.* **63** 3–11 (*203*)

Aldridge H 1986 Manoeuvrability and ecology in British bats; in *Proc 7ᵗʰ Int. Bat Res. Conf. Myotis Mitteilungsbl. Fledermauskundler* (eds.) P A Racey and H Roer, pp 157–160 (*246*)

Alexander D E 1986 Wind tunnel studies of turns by flying dragonflies; *J. Exp. Biol.* **122** 81–98 (*251*)

Alexander D E 2002 *Nature's flyers: birds, insects, and the biomechanics of flight* (Baltimore: Johns Hopkins University Press) (*111, 247*)

Alexander R M 1962 Visco-elastic properties of the body wall of sea anemones; *J. Exp. Biol.* **39** 373–386 (*204*)

Alexander R M 1966 Physical aspects of swimbladder function; *Biol. Rev.* **41** 141–176 (*158*)

Alexander R M 1969 Mechanics of the feeding action of a cyprinid fish; *J. Zool. Lond.* **159** 1–15 (*203*)

Alexander R M 1976 Estimates of speeds of dinosaurs; *Nature* **261** 129–130 (*123, 126*)

Alexander R M 1984 Stride length and speed for adults, children, and fossil hominids; *Am. J. Phys. Anthropol.* **63** 23–27 (*126*)

Alexander R M 1988a *Elastic mechanisms in animal movement* (Cambridge: Cambridge University Press) (*48, 230*)

Alexander R M 1988b Why mammals gallop; *Am. Zool.* **28** 237–245 (*128*)

Alexander R M 2000 Hovering and jumping: contrasting problems in scaling; in *Scaling in biology* (eds.) J A Brown and G B West (Oxford: Oxford University Press) pp 37–50 (*55*)

Alexander R M 2003 *Principles of animal locomotion* (Princeton, NJ: Princeton University Press) (*241*)

Alexander R M and Jayes A S 1978 Optimum walking techniques for idealized animals; *J. Zool. (London)* **186** 61–81 (*125*)

Alexander R M and Jayes A S 1983 A dynamic similarity hypothesis for the gaits of quadrupedal mammals; *J. Zool. (London)* **201** 135–152 (*123*)

Alexander R M, Jayes A S and Ker R F 1980 Estimates of energy cost for quadrupedal running gaits; *J. Zool. (London)* **190** 155–192 (*129*)

Anderson J D Jr 1997 *A history of aerodynamics* (Cambridge: Cambridge University Press) (*227*)

Arp G K and Phinney D E 1980 Ecological variations in the thermal IR emissivity of vegetation; *Envir. Exp. Bot.* **20** 135–148 (*65*)

Augspurger C K 1986 Morphology and dispersal potential of wind-dispersed diaspores of neotropical trees; *Amer. J. Bot.* **73** 353–363 (*106*)

Azuma A 1992 *The biokinetics of flying and swimming* (Tokyo: Springer-Verlag) (*106, 109, 110, 113*)

Azuma A 2006 *The biokinetics of flying and swimming, 2ⁿᵈ ed.* (Reston, VA: American Institute of Aeronautics and Astronautics) (*245*)

Azuma A and Okuno Y 1987 Flight of a samara, *Alsomitra macrocarpa*; *J. Theor. Biol.* **129** 263–274 (*111, 245*)

Baker M A and Hayward J N 1968 The influence of the nasal mucosa and the carotid rete upon hypothalamic temperature in sheep; *J. Physiol., Lond.* **198** 561–579 (*87*)

Bakken G S, Vanderbilt V C, Buttemer W A and Dawson W R 1978 Avian eggs: thermoregulatory value of very high near-infrared reflectance; *Science* **200** 321–323 (*63*)

Barnett C H, Harrison R J and Tomlinson J D W 1958 Variations in the venous system of mammals; *Biol. Rev.* **33** 442–487 (*86*)

Barthlott W and Neinhuis C 1997 Purity of the sacred lotus, or escape from contamination in biological surfaces; *Planta* **202** 1–8 (*147*)

Bartholomew G A and Lasiewski R C 1965 Heating and cooling rates, heart rate, and simulated diving in the Galapagos marine iguana; *Comp. Biochem. Physiol.* **16** 575–582 (*74*)

Bartholomew G A and Tucker V A 1964 Size, body temperature, thermal conductance, oxygen consumption, and heart rate in Australian varanid lizards; *Physiol. Zool.* **37** 341–354 (*73*)

Batham E J and Pantin C F A 1950 Muscular and hydrostatic action in the sea anemone, *Metridium senile* (L.); *J. Exp. Biol.* **27** 264–289 (*204*)

Baudinette R V 1978 Scaling of heart rate during locomotion of mammals; *J. Comp. Physiol.* **127B** 337–342 (*118*)

Bazett H C, Love L, Newton M, Eisenberg L, Day R and Forster R II 1948 Temperature changes in blood flowing in arteries and veins in man; *J. Appl. Physiol.* **1** 3–19 (*86*)

Bejan 1993 *Heat transfer* (New York: John Wiley) (*69*)

Bennet-Clark H C 1963 Negative pressures produced in the pharyngeal pump of the blood-sucking bug, *Rhodnius prolixus*; *J. Exp. Biol.* **40** 223–229 (*192, 203*)

Bennet-Clark H C 1975 The energetics of the jump of the locust *Schistocerca gregaria*; *J. Exp. Biol.* **63** 53–83 (*28, 52, 53*)

Bennet-Clark H C 1977 Scale effects in jumping animals; in *Scale effects in animal locomotion* (ed.) T J Pedley (London: Academic Press) (*40, 43*)

Bennet-Clark H C and Lucey E C A 1967 The jump of the flea: a study of the energetics and a model of the mechanism; *J. Exp. Biol.* **47** 59–76 (*29, 41, 52*)

Bennett A F, Huey R B, John-Alder H and Nagy K A 1984 The parasol tail and thermoregulatory behavior of the cape ground squirrel *Xerus inauris*; *Physiol. Zool.* **57** 57–62 (*65*)

Berg H C 1993 *Random walks in biology* (Princeton NJ: Princeton University Press) (*9*)

Bernard C 1876 *Lecons sur la Chaleur Animale* (Paris) (*84*)

Bernard R A and Wilhelm R H 1950 Turbulent diffusion in fixed beds of packed solids; *Chem. Eng. Progr.* **46** (5) 233–244 (*17*)

Bernheim, M 1959 *A sky of my own* (New York: Rinehart and Co.) (*248*)

Bidder G P 1923 The relationship of the form of a sponge to its currents; *Quart. J. Microscop. Soc.* **67** 292–323 (*203*)

Biewener A A 1990 Biomechanics of mammalian terrestrial locomotion; *Science* **250** 1097–1103 (*126*)

Biewener A A 2003 *Animal locomotion* (Oxford: Oxford University Press) (*125, 126, 128, 236*)

Biewener A, Alexander R McN and Heglund N C 1981 Elastic energy storage in the hopping of kangaroo rats (*Dipodomys spectabilis*); *J. Zool., Lond.* **195** 369–383 (*27*)

Birchall J D and Thomas N L 1983 On the architecture and function of cuttlefish bone; *J. Materials Sci.* **18** 2081–2086 (*160*)

Birkhoff G 1960 *Hydrodynamics: a study in logic, fact, and similitude 2nd ed.* (Princeton, NJ: Princeton University Press) (*105*)

Bishop W 1961 The development of tailless aircraft and flying wings; *J. Roy. Aero. Soc.* **65** 799–806 (*244*)

Bodig J and Jayne B A 1982 *Mechanics of wood and wood composites* (New York: Van Nostrand Reinhold) (*218, 219*)

Booth D T and Feder M E 1991 Formation of hypoxic boundary layers and their biological implications in a skin-breathing aquatic salamander, *Desmognathus quadramaculatus*; *Physiol. Zool.* **64** 1307–1321 (*11*)

Borrell B J 2004 Suction feeding in orchid bees (Apidae: Euglossini); *Proc. R. Soc. Lond.* **B271** (**Supp 4**) 164–166 (*198*)

Borelli G A 1680 *De motu animalium*; P Maquet, trans 1989 as *On the Movement of Animals* (Berlin: Springer-Verlag) (*39*)

Bossard R L 2002 Speed and Reynolds number of jumping cat fleas (Siphonaptera: Pulicidae); *J. Kans. Entom. Soc.* **75** 52–54 (*29*)

Brackenbury J and Hunt H 1993 Jumping in springtails: mechanism and dynamics; *J. Zool., Lond.* **229** 217–236 (*28, 34, 41*)

Brackenbury J and Wang R 1995 Ballistics and visual targeting in flea-beetles (Alticinae); *J. Exp. Biol.* **198** 1931–1942 (*28, 41, 42*)

Brainerd E L, Page B N and Fish F E 1997 Opercular jetting during fast starts by flatfishes; *J. Exp. Biol.* **200** 1179–1188 (*191*)

Bramwell C D 1971 Aerodynamics of *Pteranodon*; *Biol. J. Linn. Soc.* **3** 313–328 (*109, 110*)

Brancazio P J 1984 *Sport science: physical laws and optimum performance* (New York: Simon & Schuster) (*19, 253*)

Brown H P 1981 Key to the world genera of Larinae (Coleoptera, Dryopoidea, Elmidae), with descriptions of new genera from Hispaniola, Colombia, Australia, and New Guinea; *Pan-Pacific Entomol.* **57** 76–104 (*152*)

Brown H P 1987 Biology of riffle beetles; *Annu. Rev. Entomol.* **23** 253–273 (*145, 152*)

Bschorr O 1991 Winderregte blattschwingungen; *Naturwissenschaften* **78** 402–407 (*223*)

Buller A H R 1909 *Researches on fungi, vol. 1* (London: Longmans, Green) (*41, 51*)

Buller A H R 1933 *Researches on fungi, vol. 5* (London: Longmans, Green) (*31, 41, 51*)

Buller A H R 1934 *Researches on fungi, vol 6* (London: Longmans, Green) (*31*)

Burrows M 2006 Jumping performance of froghopper insects; *J. Exp. Biol.* **209** 4607–4621 (*28, 42*)

Bush J W M, Hu D L and Prakash M 2008 The integument of water-walking arthropods: form and function; in *Advances in insect physiology, vol 34* (eds.) J Casas and S J Simpson (London: Elsevier) pp 117–192 (*147*)

Calder W A 1984 *Size, function, and life history* (Cambridge, MA: Harvard University Press) (*118, 119*)

Campbell N A and Garber R C 1980 Vacuolar reorganization in the motor cells of *Albizzia* during leaf movements; *Planta* **148** 251–255 (*65*)

Cannell M G R and Morgan J 1987 Young's modulus of sections of living branches and tree trunks; *Tree Physiol.* **3** 355–364 (*130*)

Carey F G and Teal J M 1966 Heat conservation in tuna fish muscle; *Proc. Nat. Acad. Sci. USA* **56** 1464–1469 (*86*)

Carey F G and Teal J M 1969 Mako and porbeagle: warm-bodied sharks; *Comp. Biochem. Physiol.* **28** 199–204 (*86*)

Caro C G, Pedley T J, Schroter R C and Seed W A 1978 *The mechanics of the circulation* (Oxford: Oxford University Press) (*117*)

Carrier D R, Walter R M and Lee D V 2001 Influence of rotational inertia on turning performance of theropod dinosaurs: clues from humans with increased rotational inertia; *J. Exp. Biol.* **204** 3917–3926 (*122*)

Cavagna G A, Willems P A and Heglund N C 2000 The role of gravity in human walking: pendular energy exchange, external work, and optimal speed; *J. Physiol.* **528** 657–668 (*127*)

Caveney S, McLean H and Surry D 1998 Faecal firing in a skipper caterpillar is pressure-driven; *J. Exp. Biol.* **201** 121–133 (*28, 42*)

Cermák J, Cienciala E, Kucera J, Lindroth A and Hällgren J E 1992 Radial velocity profiles of water flow in stems of Norway spruce and oak and the response of spruce to severing; *Tree Physiol.* **10** 367–380 (*203*)

Chambers R and Hale H P 1932 The formation of ice in protoplasm; *Proc. Roy. Soc. Lond.* **B110** 336–352 (*167*)

Chappell G S 1930 *Through the alimentary canal with gun and camera: a fascinating trip to the interior* (New York: Frederick A. Stokes Co.) (*192*)

Chappell M A and Bartholomew G A 1981 Activity and thermoregulation of the antelope ground squirrel *Ammospermophilus leucurus* in winter and summer; *Physiol. Zool.* **54** 215–223 (*68*)

Chatterjee S, Templin R J and Campbell K E, Jr 2007 The aerodynamics of *Argentavis*, the world's largest flying bird from the Miocene of Argentina; *Proc. Nat. Acad. Sci USA* **104** 12398–12403 (*107, 246*)

Cheer A Y, Ogami Y and Sanderson S L 2001 Computational fluid dynamics in the oral cavity of ram suspension-feeding fishes; *J. Theor. Biol.* **210** 463–474 (*12*)

Cheng J-Y, Davison I G and DeMont M E 1996 Dynamics and energetics of scallop locomotion; *J. Exp. Biol.* **199** 1931–1946 (*203*)

Clark R B and Cowey J B 1958 Factors controlling the change of shape of certain nemertean and turbellarian worms; *J. Exp. Biol.* **35** 731–748 (*230*)

Coleman M J and Ruina A 1998 An uncontrolled walking toy that cannot stand still; *Phys. Rev. Letts.* **80** 3658–3661 (*127, 240*)

Collins S, Ruina A, Tedrake R and Wisse M 2005 Efficient bipedal robots based on passive-dynamic walkers; *Science* **307** 1082–1085 (*127, 240*)

Colmer T D and Pedersen O 2008 Underwater photosynthesis and respiration in leaves of submerged wetland plants: gas films improve CO_2 and O_2 exchange; *New Phytol.* **177** 918–926 (*148*)

Comstock J H 1887 Note on the respiration of aquatic bugs; *Am. Nat.* **21** 577–578 (*145*)

Cotterell B and Kamminga J 1990 *Mechanics of pre-industrial technology* (Cambridge: Cambridge University Press) (*238*)

Cox J P L 2008 Hydrodynamic aspects of fish olfaction; *J. R. Soc. Interface* **23** 575–593 (*14, 199*)

Crawford E C 1962 Mechanical aspects of panting in dogs; *J. Appl. Physiol.* **17** 249–251 (*75*)

Crawford E C and Kampe G 1971 Resonant panting in pigeons; *Comp. Biochem. Physiol.* **40A** 549–552 (*75*)

Crick F 1970 Diffusion in embryogenesis; *Nature* **255** 420–422 (*15*)

Cubo J and Casinos A 1998 The variation of the cross-sectional shape in the long bones of birds and mammals; *Ann. Sci. Naturelles* **1** 51–62 (*221*)

Culick F E C and Jex H R 1987 Aerodynamics, stability, and control of the 1903 Wright Flyer; in *The Wright Flyer: an engineering perspective*; (ed.) H S Wolko (Washington: Smithsonian Institution) pp 19–43 (*244*)

Curran P F 1960 Na, Cl, and water transport by rat ileum *in vitro*; *J. Gen. Physiol.* **43** 1137–1148 (*194*)

Curran P F and MacIntosh J R 1962 A model system for biological water transport; *Nature* **193** 347–348 (*194*)

Dalton S 2001 *The miracle of flight* (London: Merrell Publishers Ltd) (*243*)

Daniel T L 1984 Unsteady aspects of aquatic locomotion; *Amer. Zool.* **24** 121–134 (*105*)

Daniel T L and Kingsolver 1983 Feeding strategy and the mechanics of blood sucking in insects; *J. Theor. Biol.* **105** 661–672 (*203*)

Darwin C 1845 *Journal of researches into the natural history and geology of the countries visited during the voyage of H.M.S. Beagle under the command of Capt. Fitz Roy, R.N.* (London: J Murray) (*74*)

Dawson T H 2005 Modeling of vascular networks; *J. Exp. Biol.* **208** 1687–1694 (*118*)

Deban S M, Wake D B and Roth G 1997 Salamander with a ballistic tongue; *Nature* **389** 27–28 (*28*)

Dechow P C and Hylander W L 2000 Elastic properties and masticatory bone stress in the macaque mandible; *Am. J. Phys. Anthropol.* **112** 553–574 (*222*)

DeMont M E and Gosline J M 1988 Mechanics of jet propulsion in the hydromedusan jellyfish, *Polyorchis penicillatus*. II. Energetics of the jet cycle; *J. Exp. Biol.* **134** 333–345 (*191, 203*)

Denny M W 1988 *Biology and the mechanics of the wave-swept environment* (Princeton, NJ: Princeton University Press) (*105*)

Denny M W 1993 *Air and water: the biology and physics of life's media* (Princeton: Princeton University Press) (*104*)

Denny M W and Gaines S 2000 *Chance in biology: using probability to explore nature* (Princeton, NJ: Princeton University Press) (*88*)

Denton E J and Gilpin-Brown J B 1961 The buoyancy of the cuttlefish, *Sepia officinalis* (L.); *J. Mar. Biol. Assoc. UK* **41** 319–342 (*160, 161*)

Denton E J, Gilpin-Brown J B and Howarth J V 1961 The osmotic mechanism of the cuttlebone; *J. Mar. Biol. Assoc. UK* **41** 351–364 (*160*)

DeVries A L 1983 Antifreeze peptides and glycopeptides in cold-water fishes; *Annu. Rev. Physiol.* **45** 245–260 (*167*)

DeVries A L and Eastman J T 1978 Lipid sacs as a buoyancy adaptation in an Antarctic fish; *Nature* **271** 352–353 (*156*)

Dewar H, Graham J B and Brill R W 1994 Studies of tropical tuna swimming performance in a large water tunnel. II. Thermoregulation; *J. Exp. Biol.* **192** 33–44 (*86*)

Diamond J M and Bossert W H 1967 Standing-gradient osmotic flow; *J. Gen. Physiol.* **50** 2061–2082 (*194*)

Dietz C, Dietz I and Siemers B M 2006 Wing measurement variation in the five European horseshoe bat species (Chiroptera: Rhinolophidae); *J. Mammal.* **87** 1241–1251 (*246*)

Disney W 1953 *The living desert* (Vol. 2 in *Walt Disney's legacy collection*, DVD 2006) (*257*)

Drost M R, Muller M and Osse J W M 1988 A quantitative hydrodynamical model of suction feeding in larval fishes: the role of frictional forces; *Proc. R. Soc. Lond.* **B234** 263–281 (*203*)

DuBois E F 1939 Heat loss from the human body; *Bull. N.Y. Acad. Med.* **13** 143–173 (*82*)

Dudley R 2000 *The biomechanics of insect flight* (Princeton, NJ: Princeton University Press) (*105, 113, 114, 247*)

Dudley R and DeVries P 1990 Tropical rain forest structure and the geographical distribution of gliding vertebrates; *Biotropica* **22** 432–434 (*111*)

Dudley R and Ellington C P 1990 Mechanics of forward flight in bumblebees. II. Quasi-steady lift and power calculations; *J. Exp. Biol.* **148** 53–88 (*114*)

Dunkin R C, McLellan W A, Blum J E and Pabst D A 2005 The ontogenetic changes in the thermal properties of blubber from Atlantic bottlenose dolphin *Tursiops truncatus*; *J. Exp. Biol.* **208** 1469–1480 (*77, 81*)

Eastman J T and DeVries A L 1982 Buoyancy studies of notothenioid fishes in McMurdo Sound, Antarctica; *Copeia* **1982** 385–393 (*156*)

Edney E B 1977 *Water balance in land arthropods* (Berlin: Springer-Verlag) (*177*)

Edsall J T and Wyman J 1958 *Biophysical chemistry* (New York: Academic Press) (*169*)

Edwards M H 1986 Zero angular momentum turns; *Am. J. Phys.* **54** 846–847 (*253, 254, 255*)

Ellers O 1995 Form and motion of *Donax variabilis* in flow; *Biol. Bull.* **189** 138–147 *(155)*

Ellington C P 1984 The aerodynamics of hovering flight; *Phil. Trans. Roy. Soc. Lond.* **B305** 1–181 *(114)*

Ellington C P 1991 Aerodynamics and the origin of insect flight; *Adv. Insect Physiol.* **23** 171–210 *(247)*

Ellmore G S and Ewers F W 1986 Fluid flow in the outermost xylem increment of a ring-porous tree, *Ulmus americana*; *Am. J. Bot.* **73** 1771–1774 *(133)*

Ennos A R 1988 The importance of torsion in the design of insect wings; *J. Exp. Biol.* **140** 137–160 *(227)*

Ennos A R 1993a. The function and formation of buttresses; *TREE* **8** 350–351 *(137)*

Ennos A R 1993b. The mechanics of the flower stem of the sedge *Carex acutiformis*; *Ann. Bot.* **72** 123–127 *(224)*

Ennos A R 1995 Mechanical behaviour in torsion of insect wings, blades of grass and other cambered structures; *Proc. R. Soc. London* **B259** 15–18 *(227)*

Ennos A R 1999 The aerodynamics and hydrodynamics of plants; *J. Exp. Biol.* **202** 3281–3284 *(138)*

Ennos A R, Spatz H-Ch and Speck T 2000 The functional morphology of the petioles of the banana, *Musa textilis*; *J. Exp. Bot.* **51** 2085–2093 *(224)*

Essner R L 2002 Three-dimensional launch kinematics in leaping, parachuting and gliding squirrels; *J. Exp. Biol.* **205** 2469–2477 *(43)*

Etnier S A 2001 Flexural and torsional stiffness in multi-jointed biological beams; *Biol. Bull.* **200** 1–8 *(228)*

Etnier S A 2003 Twisting and bending of biological beams: distribution of biological beams in a stiffness mechanospace; *Biol Bull.* **205** 36–46 *(216)*

Etnier S A and Vogel S 2000 Reorientation of daffodil (*Narcissus*) flowers in wind: drag reduction and torsional flexibility; *Am. J. Bot.* **07** 29–32 *(224, 229)*

Evans M E G 1972 The jump of the click beetle (Coleoptera, Elateridae): a preliminary study; *J. Zool., Lond.* **167** 319–336 *(28, 42)*

Evans M E G 1975 The jump of *Petrobius* (Thysanura, Machilidae); *J. Zool., Lond.* **176** 49–65 *(28, 42)*

Fänge R 1983 Gas exchange in fish swim bladder; *Rev. Physiol. Biochem. Pharmacol.* **97** 111–158 *(158, 159)*

Farley C T, Houdijk H H P, Van Strien C and Louie M 1998 Mechanism of leg stiffness adjustment for hopping on surfaces of different stiffnesses; *J. Appl. Physiol.* **85** 1044–1055 *(228)*

Federspiel W J and Fredberg J J 1988 Axial dispersion in respiratory bronchioles and alveolar ducts; *J. Appl. Physiol.* **64** 2614–2621 *(15)*

Feng L, Li S, Li Y, Li H, Zhang L, Zhai J, Song Y, Liu B, Jiang L and Zhu D 2002 Super-hydrophobic surfaces: from natural to artificial; *Advanced Mat.* **14** 1857–1860 *(147)*

Ferris D P and Farley C T 1997 Interaction of leg stiffness and surface stiffness during human hopping; *J. Appl. Physiol.* **82** 15–22 *(228)*

Fichtner K and Schulze E-D 1990 Xylem water flow in tropical vines as measured by a steady state heating method; *Oecologia* **82** 355–361 *(203)*

Fischer M, Cox J, Davis D J, Wagner A, Taylor R, Huerta A J and Money N P 2004 New information on the mechanism of forcible ascospore discharge from *Ascobolus immersus*; *Fungal Genet. Biol.* **41** 698–707 (*32, 41*)

Fish F E 1979 Thermoregulation in the muskrat (*Ondatra zibethicus*): the use of regional heterothermia; *Comp. Biochem. Physiol.* **64A** 391–397 (*86*)

Fish F E 2002 Balancing requirements for stability and maneuverability in cetaceans; *Integ. Comp. Biol.* **42** 85–93 (*250, 251*)

Fish F E and Full R J 2002 Stability and maneuverability; *Integ. Comp. Biol.* **42** 85–181 (*232*)

Fish F E, Nicastro A J and Weihs D 2006 Dynamics of the aerial maneuvers of spinner dolphins; *J. Exp. Biol.* **209** 590–598 (*256*)

Fish F E and Stein B R 1991 Functional correlates of differences in bone density among terrestrial and aquatic genera in the family Mustelidae (Mammalia); *Zoomorphology* **110** 339–345 (*155*)

Fletcher G L, Choy L H and Davies P L 2001 Antifreeze proteins of teleost fishes; *Annu. Rev. Physiol.* **63** 359–390 (*167*)

Flynn M R and Bush J W M 2008 Underwater breathing: the mechanics of plastron respiration; *J. Fluid Mech.* **608** 275–296 (*147, 148*)

Forouhar A S, Liebling M, Hickerson A, Nasiraei-Moghaddam A, Tsai H-J, Hove J R, Fraser S E, Dickinson M E and Gharib M 2006 The embryonic vertebrate heart is a dynamic suction tube; *Science* **312** 751–753 (*206*)

Forterre Y, Skotheim J M, Dumais J and Mahadevan L 2005 How the Venus flytrap snaps; *Nature* **433** 421–425 (*51, 193*)

Foster-Smith R L 1978 An analysis of water flow in tube-living animals; *J. Exp. Mar. Biol. Ecol.* **34** 73–95 (*195, 203*)

Frolich C 1979 Do springboard divers violate angular momentum conservation? *Am. J. Phys.* **47** 583–592 (*252, 253*)

Full R J, Kubow T, Schmitt J, Holmes P and Koditschek D 2002 Quantifying dynamic stability and maneuverability of terrestrial vertebrates; *Integ. Comp. Biol.* **42** 149–157 (*241*)

Gallenmüller F, Rowe N and Speck T 2004 Development and growth form of the neotropical liana *Croton nuntians*: the effect of light and mode of attachment on the biomechanics of the stem; *J. Plant Growth Regul.* **23** 83–87 (*220*)

Gao X and Jiang L 2004 Water-repellent legs of water striders; *Nature* **432** 36 (*147*)

Garrison W J, Miller G L and Raspet R 2000 Ballistic seed projection in two herbaceous species; *Amer. J. Bot.* **87** 1257–1264 (*30, 42*)

Gartner B L 1995 Patterns of xylem variation within a tree and their hydraulic and mechanical consequences in *Plant stems: physiology and functional morphology* (ed.) B L Gartner (San Diego: Academic Press) pp 125–149 (*134*)

Gates D M 1965 Energy, plants, and ecology; *Ecology* **46** 1–13 (*62*)

Gates D M 1980 *Biophysical ecology* (New York: Springer-Verlag) (*62, 76*)

Geiger R 1965 *The climate near the ground, rev. ed.* (Cambridge, MA: Harvard University Press) (*101*)

Gerwick W H and Lang N J 1977 Structural, chemical and ecological studies on iridescence in *Iridaea* (Rhodophyta); *J. Phycol.* **13** 121–127 (*64*)

Gibbons C A and Shadwick R E 1991 Circulatory mechanics in the toad *Bufo marinus*. II. Hemodyamics of the arterial windkessel; *J. Exp. Biol.* **158** 291–306 (*203*)

Gibo D L and Pallett M J 1979 Soaring flight of monarch butterflies; *Can. J. Zool.* **57** 1393–1401 (*110*)

Giller P S and Malmqvist B 1998 *The biology of streams and rivers* (Oxford: Oxford University Press) (*152*)

Givnish T J 1995 Plant stems: biomechanical adaptations for energy capture and influence on species distributions in *Plant stems: physiology and functional morphology* (ed.) B L Gartner (San Diego: Academic Press) pp 3–49 (*139*)

Glad Sorensen H, Lambrechtsen J and Einer-Jensen N 1991 Efficiency of the counter current transfer of heat and ^{133}xenon between the pampiniform plexis and testicular artery of the bull; *Int. J. Androl.* **14** 232–240 (*86*)

Glasheen J W and McMahon T A 1996 Size-dependence of water-running ability in basilisk lizards (*Basiliscus basiliscus*); *J. Exp. Biol.* **199** 2611–2618 (*98*)

Glemain J, Cordonnier J P, LeNormand L and Buzelin J M 1990 Urodynamic consequences of urethral stenosis: hydrodynamic study with a theoretical model; *J. d'Urol.* **96** 271–277 (*191, 203*)

Goddard C K 1972 Structure of the heart of the ascidian *Pyura praeputialis*; *Aust. J. Biol. Sci.* **25** 645–647 (*206*)

Gordon J E 1978 *Structures, or why things don't fall down* (London: Penguin Books) (*227*)

Gosline J M 1971 Connective tissue mechanics of *Metridium senile*. II. Viscoelastic properties and a macromolecular model; *J. Exp. Biol.* **55** 775–795 (*204*)

Gosline J, Lillie M, Carrington E, Guerette P, Ortlepp C and Savage K 2002 Elastic proteins: biological roles and mechanical properties; *Phil. Trans. R. Soc. Lond.* **B357** 121–132 (*53*)

Gould S J 1981 Kingdoms without wheels; *Nat. Hist.* **90(4)** 42–48 (*122*)

Gower D and Vincent J F V 1996 The mechanical design of the cuttlebone and its bathymetric implications; *Biomimetics* **4** 37–57 (*160*)

Greenhill A G 1881 Determination of the greatest height consistent with stability that a vertical pole or mast can be made, and of the greatest height to which a tree of given proportions can grow; *Cambridge Phil. Soc.* **4** 65–73 (*130, 131*)

Greer A E, Lazell J D and Wright R M 1973 Anatomical evidence for a countercurrent heat exchanger in the leatherback turtle (*Dermochelys coriacea*); *Nature* **244** 181 (*86*)

Griffin T M and Kram R 2000 Penguin waddling is not wasteful; *Nature (London)* **408** 929 (*127, 239*)

Griffin T M, Tolani N A and Kram R 1999 Walking in simulated reduced gravity: mechanical energy fluctuations and exchange; *J. Appl. Physiol.* **86** 383–390 (*127*)

Grubb B 1983 Allometric relations of cardiovascular function in birds; *Am. J. Physiol.* **245** H567–H572 (*118*)

Gunga H C, Kirsch, K A, Baartz F, Röcker L, Heinrich W–D, Lisowski W, Wiedemann A and Albertz J 1995 New Data on the dimensions of *Brachiosaurus brancai* and their physiological implications; *Naturwissenschaften* **82** 190–192 (*122*)

Guy C L 1990 Cold acclimation and freezing stress tolerance: role of protein metabolism; *Annu. Rev. Plant Physiol. Plant Mol. Biol.* **41** 187–233 (*168*)

Hadley N F 1994 *Water relations of terrestrial arthropods* (San Diego: Academic Press) (*76, 77, 176, 177*)

Haldane J B S 1926 On being the right size; *Harper's Monthly* **152** 424–427 (*39*)

Haldane J S 1922 *Respiration* (New Haven, CT: Yale University Press) (*84*)

Hall F G, Dill D B and Barron E S G 1936 Comparative physiology in high altitudes; *J. Cell. Comp. Physiol.* **8** 301–313 (*100*)

Harpster H T 1941 An investigation of the gaseous plastron as a respiratory mechanism in *Helichus striatus* Leconte (Dryopidae); *Trans. Am. Microscop. Soc.* **60** 329–358 (*145, 148*)

Harpster H T 1944 The gaseous plastron as a respiratory mechanism in *Stenelmis quadrimaculata* Horn (Dryopidae); *Trans. Am. Microscop. Soc.* **63** 1–26 (*145*)

Harris J S 1989 An airplane is not a bird; *Amer. Heritage of Invention and Technology* **5**(2) 19–22 (*243*)

Hasan Z 2005 The human motor control system's response to mechanical perturbation: should it, can it, and does it ensure stability? *J. Motor Behav.* **37** 484–493 (*235*)

Hatch D E 1970 Energy conserving and heat dissipating mechanisms of the turkey vulture; *Auk* **87** 111–124 (*75*)

Hayami I 1991 Living and fossil scallop shells as airfoils: an experimental study; *Paleobiology* **17** 271–276 (*109*)

Hebets E H and Chapman R F 2000 Surviving the flood: plastron respiration in the non-tracheate arthropod *Phrynus marginemaculatus* (Amblypygi: Arachnida); *J. Insect Physiol.* **46** 13–19 (*147*)

Hedrich T L, Usherwood J R and Biewener A A 2007 Low speed maneuvering flight of the rose-breasted cockatoo (*Eolophus roseicapillus*). II. Inertial and aerodynamic reorientation; *J. Exp. Biol.* **210** 1912–1934 (*256*)

Heglund N C and Taylor C R 1988 Speed, stride frequency and energy cost per stride: how do they change with body size and gait? *J. Exp. Biol.* **138** 301–318 (*129*)

Heglund N C, Taylor C R and McMahon T A 1974 Scaling stride frequency and gait to animal size: mice to horses; *Science* **186** 1112–1113 (*129*)

Heinrich B 1996 *The thermal warriors: strategies of insect survival* (Cambridge, MA: Harvard University Press) (*65, 76, 87*)

Henderson L J 1913 *The fitness of the environment* (New York: The Macmillan Co.) (*164*)

Hennemann W W III 1982 Energetics and spread-winged behavior of anhingas in Florida; *Condor* **84** 91–92 (*158*)

Hertel H 1966 *Structure, form and movement* (New York: Reinhold) (*196*)

Hill A V 1938 The heat of shortening and dynamic constants of muscle; *Proc. R. Soc. Lond.* **B126** 136–195 (*39*)

Hill D R 1973 Trebuchets; *Viator* **4** 99–114 (*47*)

Hills G J 1972 The physics and chemistry of high pressures; *Symp. Soc. Exp. Biol.* **26** 1–26 (*142*)

Hinton H E 1976 Plastron respiration in bugs and beetles; *J. Insect Physiol.* **22** 1529–1550 (*146, 147*)

Hoffmann B, Chabbert B, Monties B and Speck T 2003 Mechanical, chemical and X-ray analysis of wood in the two tropical lianas *Bauhinia guianensis* and *Condylocarpon guianense*: variations during ontogeny; *Planta* **217** 32–40 (*218, 220*)

Holbrook N M and Zwieniecki M A 1999 Embolism repair and xylem tension: do we need a miracle? *Plant Physiol.* **120** 7–10 (*133*)

Holstein T and Tardent P 1984 An ultrahigh-speed analysis of exocytosis: nematocyst discharge; *Science* **223** 830–832 (*37*)

Hondzo M and Lyn D 1999 Quantified small-scale turbulence inhibits the growth of a green alga; *Freshwater Biol.* **41** 51–61 (*14*)

Hora S L 1930 Ecology, bionomics and evolution of the torrential fauna, with special reference to the organs of attachment; *Phil. Trans. R. Soc.* **B218** 171–182 (*152*)

Howland H C and Howland B 1962 The reaction of blinded goldfish to rotation in a centrifuge; *J. Exp. Biol.* **39** 491–502 (*249*)

Hu D L, Prakash M, Chan B and Bush J W M 2007 Water-walking devices; *Exptl. Fluids* **43** 769–778 (*98*)

Hughes G M 1958 The co-ordination of insect movements; *J. Exp. Biol.* **35** 567–583 (*191*)

Hughes G M 1966 The dimensions of fish gills in relation to their function; *J. Exp. Biol.* **45** 177–195 (*11*)

Humphrey J A C and Barth F G 2008 Medium flow-sensing hairs: biomechanics and models; in *Advances in insect physiology, vol 34* (eds.) J Casas and S J Simpson (London: Elsevier) pp 1–80 (*249*)

Humphreys W F 1975 The influence of burrowing and thermoregulatory behaviour on the water relations of *Geolycosa godeffroyi* (Araneae: Lycosidae), an Australian wolf spider; *Oecologia* **21** 291–311 (*179*)

Hunter T and Vogel S 1986 Spinning embryos enhance diffusion through gelatinous egg masses; *J. Exp. Mar. Biol. Ecol.* **96** 303–308 (*67*)

Hutchinson J R and Garcia M 2002 *Tyrannosaurus* was not a fast runner; *Nature (London)* **415** 1018–1021 (*126*)

Ingold C T 1939 *Spore Discharge in Land Plants* (Oxford: Clarendon Press) (*32, 34*)

Ingold C T 1971 *Fungal spores: their liberation and dispersal* (Oxford: Clarendon Press) (*104, 110*)

Ingold C T and Hadland S A 1959 The ballistics of *Sordaria*; *New Phytol.* **58** 46–57 (*32, 41, 51*)

Irving L and Hart J S 1957 The metabolism and insulation of seals as bare-skinned mammals in cold water; *Can. J. Zool.* **35** 497–511 (*82*)

Irving L and Krog J 1955 Temperature of the skin in the Arctic as a regulator of heat; *J. Appl. Physiol.* **7** 355–364 (*86*)

Jackson D C and Schmidt-Nielsen K 1964 Countercurrent heat exchange in the respiratory passages; *Proc. Nat. Acad. Sci. USA* **51** 1192–2297 (*87, 174*)

Jackson R D and van Bavel C H M 1965 Solar distillation of water from soil and plant materials: a simple desert survival technique; *Science* **149** 1377–1379 (*177, 178*)

Jayes A S and Alexander R McN 1980 The gaits of chelonians: walking techniques for very low speeds; *J. Zool., Lond.* **191** 353–378 (*241*)

Jensen M 1956 Biology and physics of locust flight. III. The aerodynamics of locust flight; *Phil. Trans. Roy. Soc. Lond.* **B239** 511–552 (*110*)

Jensen M and Weis-Fogh T 1962 Biology and physics of locust flight. V. Strength and elasticity of locust cuticle; *Phil. Trans. R. Soc. Lond.* **B245** 137–169 (*53*)

Jeyasuria P and Lewis J C 1987 Mechanical properties of the axial skeleton in gorgonians; *Coral Reefs* **5** 213–219 (*225*)

Johansen K and Martin A W 1965 Circulation in a giant earthworm, *Glossoscolex giganteus*. I. Contractile processes and pressure gradients in the large vessels; *J. Exp. Biol.* **43** 333–347 (*191*)

Johnson P S and Shifley S R 2002 *The ecology and silviculture of oaks* (New York: CABI Publishing) (*132*)

Jones H D 1983 Circulatory systems of gastropods and bivalves; in *The mollusca, v5* (ed.) K M Wilbur (New York: Academic Press) pp 189–238 (*203*)

Jones J C 1977 *The circulatory system of insects* (Springfield, IL: Charles C. Thomas) (*191, 192*)

Jones J M, Wentzell L A and Toews D P 1992 Posterior lymph heart pressure and rate and lymph flow in the toad *Bufo marinus* in response to hydrated and dehydrated conditions; *J. Exp. Biol.* **169** 207–220 (*203*)

Jusufi A, Goldman D I, Revzen S and Full R J 2008 Active tails enhance arboreal acrobatics in geckos; *Proc. Nat. Acad. Sci. USA* **105** 4215–4219 (*256*)

Kaandorp J A and Sloot P M A 2001 Morphological models of radiative accretive growth and the influence of hydrodynamics; *J. Theor. Biol.* **209** 257–274 (*13*)

Kane T R and Scher M P 1969 A dynamical explanation of the falling cat phenomenon; *Int. J. Solids Structures* **5** 663–670 (*254*)

Kant Y and Badarinath K V S 2002 Ground-based method for measuring thermal infrared effective emissivities: implications and perspectives on the measurement of land surface temperature from satellite data; *Int. J. Remote Sensing* **23** 2179–2191 (*66*)

Kanwisher J W 1955 Freezing in intertidal animals; *Biol Bull.* **109** 56–63 (*168*)

Kanwisher J W 1959 Histology and metabolism of frozen intertidal animals; *Biol. Bull.* **116** 258–264 (*168*)

Karassik I J, Messina J P, Cooper P and Heald C C 2000 *Pump handbook, 3rd ed.* (New York: McGraw Hill Professional) (*187*)

Kardong K and Lavin-Murcio P 1993 Venom delivery of snakes as high-pressure and low-pressure systems; *Copeia* **1993** 644–650 (*191, 203*)

Katz S L and Gosline J M 1993 Ontogenetic scaling of jump performance in the African desert locust (*Schistocerca gregaria*); *J. Exp. Biol.* **177** 81–111 (*42, 43*)

Kelly D A 2002 The functional morphology of penile erection: tissue designs for increasing and maintaining stiffness; *Integr. Comp. Biol.* **42** 216–221 (*230*)

Kilgore D L Jr and Schmidt-Nielsen 1975 Heat loss from ducks' feet immersed in cold water; *Condor* **77** 475–478 (*86*)

Kim T K, Silk W K and Cheer A Y 1999 A mathematical model for pH patterns in the rhizospheres of growth zones; *Plant, Cell and Environment* **22** 1527–1538 (*10*)

Kincaid, D T 1976 *Theoretical and experimental investigations of* Ilex *pollen and leaves in relation to microhabitat in the southeastern United States* PhD thesis, Wake Forest University, Winston-Salem, NC. (*91*)

Kingsolver J G and Daniel T L 1983 Mechanical determinants of nectar feeding in hummingbirds: energetics, tongue morphology, and licking behavior; *Oecologia* 60 214–226 (*198, 203*)

Kingsolver J G and Daniel T L 1995 Mechanics of food handling by fluid-feeding insects; in *Regulatory mechanisms in insect feeding* (eds.) R F Chapman and G deBoer (New York: Chapman and Hall) pp 32–73 (*191, 198*)

Klein, D R 2001 Similarity in habitat adaptations of arctic and African ungulates: evolutionary convergence or ecological divergence? *Alces* 37 245–252 (*175*)

Kling G W, Clark M A, Compton H R, Devine J D, Evans W C, Humphrey A M, Koenigsberg E J, Lockwood J P, Tuttle M L and Wagner G L 1987 The 1986 Lake Nyos gas disaster in Cameroon, West Africa; *Science* 236 169–175 (*101*)

Koch G W, Sillett S C, Jennings G M and Davis S D 2004 The limit to tree height; *Nature* 428 851–854 (*135*)

Koehl M A R 1977 Mechanical diversity of connective tissue of the body wall of sea anemones; *J. Exp. Biol.* 69 107–125 (*204*)

Koehl M A R, Jumars P A, and Karp-Boss L 2003 Algal biophysics; in *Out of the past* (ed.) T A Norton (Belfast: British Phycological Association) pp 115–130 (*105*)

Kramer P J 1959 Transpiration and the water economy of plants; in *Plant physiology v2* (ed.) F C Steward (New York: Academic Press) pp 607–726 (*203*)

Krisper G 1990 Das Sprungvermögen der mitbengattung *Zetorchestes* (Acarida, Oribatida); *Zool. Jb. Anat.* 120 289–312 (*41*)

Krogh A 1941 *The comparative physiology of respiratory mechanisms* (Philadelphia: University of Pennsylvania Press) (*71, 145, 169*)

Kruur J, Brailsford L C, Glofcheski D J and Lepock J R 1985 Effect of dissolved gases on freeze-thaw survival of mammalian cells; *Cryo-letters* 6 233–238 (*169, 172*)

LaBarbera M 1983 Why the wheels won't go; *Am. Nat.* 121 395–408 (*122*)

LaBarbera M 1990 Principles of design of fluid transport systems in zoology; *Science* 249 992–1000 (*4*)

LaBarbera M and Vogel S 1982 The design of fluid transport systems in organisms; *Am. Sci.* 70 54–60 (*197*)

Lai N, Shabetai R, Graham J B, Hoit B D, Sunnerhagen K S and Bhargava V 1990 Cardiac function of the leopard shark, *Triakis semifasciata*; *J. Comp. Physiol.* B160 259–268 (*203*)

Lancashire J R and Ennos A R 2002 Modelling the hydrodynamic resistance of bordered pits; *J. Exp. Bot.* 53 1485–1493 (*134*)

Lapennas G N and Schmidt-Nielsen K 1977 Swimbladder permeability to oxygen; *J. Exp. Biol.* 67 175–196 (*159*)

Lasiewski R C and Bartholomew G A 1969 Condensation as a mechanism for water gain in nocturnal desert poikilotherms; *Copeia* 1969 405–407 (*181*)

Lauder G V 1980 The suction feeding mechanism in sunfishes (*Lepomis*): an experimental analysis; *J. Exp. Biol.* 88 49–72 (*191*)

Lauder G V 1984 Pressure and water flow patterns in the respiratory tract of the bass (*Micropterus salmoides*); *J. Exp. Biol.* **113** 151–164 (*203*)

Lee R E and Costanzo J P 1998 Biological ice nucleation and ice distribution in cold-hardy ectothermic animals; *Annu. Rev. Physiol.* **60** 55–72 (*166, 167*)

Levy M and Salvadori M 2002 *Why buildings fall down, updated and expanded* (New York: W W Norton) (*230*)

Lilienthal O 1889 *Der Vogelflug als Grundlage der Fliegekunst.* Trans. 1910 as *Bird flight as the basis for aviation* (London: Longmans, Green) (*243*)

Lillie M and Boechler T 2005 (formally unpublished and untitled; see http://web.archive.org/web/20050320145500/http://www.zoology.ubc.ca/~biol438/Reports/CowTip.PDF). (*238*)

Lillywhite H B 2006 Water relations of tetrapod integument; *J. Exp. Biol.* **209** 202–226 (*198*)

Lillywhite H B and Licht P 1974 Movement of water over toad skin: functional role of epidermal sculpturing; *Copeia* **1974** 165–171 (*198*)

Lim A T 1969 *A study of convective heat transfer from plant leaf models* MS thesis, Duke University, Durham, NC (*70*)

Linderstrøm-Lang K 1937 Principle of the Cartesian diver applied to gasometric technique; *Nature* **140** 1080 (*161, 162*)

Lipp G, Körber C, Englich S, Hartmann U and Rau G 1987 Investigation of the behavior of dissolved gases during freezing; *Cryobiology* **24** 489–503 (*169, 172*)

Lloyd J T 1914 Lepidopterous larvae from rapid streams; *J. N.Y. Entomol. Soc.* **22** 145–152 (*154*)

Lovvorn J R and Jones D R 1991 Effects of body size, body fat, and changes in pressure with depth on buoyancy and cost of diving in ducks (*Aythya* spp.); *Can. J. Zool.* **69** 2879–2887 (*158*)

MacIntyre S 1993 Vertical mixing in a shallow, eutrophic lake: possible consequences for the light climate of phytoplankton; *Limnol. Oceanogr.* **38** 798–817 (*9*)

Maherali H, Moura C F, Caldeira M C, Willson C J and Jackson R B 2006 Functional coordination between leaf gas exchange and vulnerability to xylem cavitation in temperate forest trees; *Plant, Cell and Environ.* **29** 571–583 (*133*)

Maier K (1974) Ruptur der Kupsalwand bei *Sphagnum*; *Plant Syst. Evol.* **123** 13–24 (*32*)

Maitland D P 1992 Locomotion by jumping in the Mediterranean fruit-fly larva *Ceratitis capitata*; *Nature* **355** 159–161 (*28, 42*)

Marden J H and L R Allen 2002 Universal performance characteristics of motors; *Proc. Nat. Acad. Sci. USA* **99** 4161–4166 (*55*)

Mares M A 2002 *A desert calling: life in a forbidding landscape* (Cambridge, MA: Harvard University Press) (*257*)

Maresh J L, Fish F E, Nowacek D P, Nowacek S M, and Wells R S 2004 High performance turning capabilities during foraging by bottlenose dolphins (Tursiops truncatus); *Marine Mammal Sci.* **203** 498–509 (*256*)

Marsh R L 1994 Jumping ability of anuran amphibians; *Adv. Vet. Sci.* **38B** 51–111 (*43, 48*)

Marsh R L and John-Alder H B 1994 Jumping performance of hylid frogs measured with high-speed cine film; *J. Exp. Biol.* **188** 131–141 (*43*)

Martel Y 2001 *The life of pi* (Toronto: Alfred A. Knopf Canada) (*178*)

Martin A W 1974 Circulation in invertebrates; *Annu. Rev. Physiol.* **36** 171–186 (*192, 206*)

Martin A W, F M Harrison, Huston M J and Stewart D W 1958 The blood volume of some representative molluscs; *J. Exp. Biol.* **35** 260–279 (*4*)

Maynard Smith J 1952 The importance of the nervous system in the evolution of animal flight; *Evolution* **6** 127–129 (*242*)

Mayr S, Gruber A and Bauer H 2003 Repeated freeze-thaw cycles induce embolism in drought stressed conifers (Norway spruce, stone pine); *Planta* **217** 436–441 (*169*)

McGahan J 1973 Gliding flight of the Andean condor in nature; *J. Exp. Biol.* **58** 225–237 (*110*)

McGuire J A and Dudley R 2005 Comparative gliding performance of flying lizards; *Amer. Nat.* **166** 93–106 (*109, 111*)

McMahon T A 1973 Size and shape in biology; *Science* **179** 1201–1202 (*132*)

McMahon T A 1984 *Muscles, reflexes, and locomotion* (Princeton, NJ: Princeton University Press) (*46*)

McMahon T A and Bonner J T 1983 *On size and life* (New York: Scientific American Library) (*113*)

McQueen D J and Culik B 1981 Field and laboratory activity patterns in the burrowing wolf spider *Geolycosa domifex* (Hancock); *Can. J. Zool.* **59**: 1263–1271 (*181*)

Mead K S 2005 Sensory biology: linking the internal and external ecologies of marine organisms; *Mar. Ecol. Prog. Ser.* **287** 285–289 (*14*)

Mead K S and Weatherby T M 2002 Morphology of stomatopod chemosensory sensilla facilitates fluid sampling; *Invert. Biol.* **121** 148–157 (*14*)

Merks R, Hoekstra A, Kaandorp J and Sloot P 2003 Models of coral growth: spontaneous branching, compactification and the Laplacian growth assumption; *J. Theor. Biol.* **224** 153–166 (*13*)

Michaelides E E 1997 Review—the transient equation of motion for particles, bubbles, and droplets; *J. Fluids Engin.* **119** 233–247 (*105*)

Middleman S 1972 *Transport phenomena in the cardiovascular system* (New York: John Wiley) (*4*)

Milburn J A 1979 *Water flow in plants* (London: Longmans) (*134*)

Miller L A 2005 Structural dynamics and resonance in plants with non-linear stiffness; *J. Theor. Biol.* **234** 511–524 (*228*)

Milnor W R 1990 *Cardiovascular physiology* (New York: Oxford University Press) (*203*)

Minetti A E 1995 Optimum gradient of mountain paths; *J. Appl. Physiol.* **79** 1698–1703 (*127*)

Minetti A E 2001 Walking on other planets; *Nature (London)* **409** 467–468 (*127*)

Mitchell J G, Pearson L, Bonazinga A, Dillon S, Khouri H and Paxinos R 1995 Long lag times and high velocities in the motility of natural assemblages of marine bacteria. *Appl. Environ. Microbiol.* **61** 877–882 (*9*)

Monteith J L and Unsworth M 1990 *Principles of environmental physics, 2nd ed.* (Oxford: Butterworth-Heinemann) (*62, 90, 104*)

Mooney H A, Gulmon S L, Ehleringer J and Rundel P W 1980 Atmospheric water uptake by an Atacama desert shrub; *Science* **209** 693–694 (*176*)

Müller J 1833 On the existence of four distinct hearts, having regular pulsations, connected with the lymphatic system, in certain amphibious animals; *Phil. Trans. Roy. Soc. Lond.* **123** 89–94 (*203*)

Munk W H and Riley G A 1952 Absorption of nutrients by aquatic plants; *J. Mar. Res.* **11** 215–240 (*8*)

Murphy D J 1983 Freezing resistance in intertidal invertebrates; *Annu. Rev. Physiol.* **45** 289–299 (*168, 172*)

Neinhuis C and Barthlott W 1997 Characterization and distribution of water-repellent, self-cleaning plant surfaces; *Ann. Bot.* **79** 667–677 (*147*)

Nielsen A 1950 The torrential invertebrate fauna; *Oikos* **2** 176–196 (*152, 153*)

Nielsen N F, Larson P S, Riisgård H U and Jørgensen C B 1993 Fluid motion and particle retention in the gill of *Mytilus edulis*: Video recordings and numerical modelling; *Mar. Biol.* **116** 61–71 (*11*)

Niklas K J 1984 The motion of windborne pollen grains around conifer ovulate cones: implications on wind pollination; *Amer. J. Bot.* **71** 356–374 (*110*)

Niklas K J 1991 The elastic moduli and mechanics of *Populus tremuloides* (Salicaceae) petioles in bending and torsion; *Am. J. Bot.* **78** 989–996 (*223*)

Niklas K J 1992 *Plant biomechanics: an engineering approach to plant form and function* (Chicago: University of Chicago Press) (*132, 230*)

Niklas K J 1997a. Relative resistance of hollow, septate internodes to twisting and bending; *Ann. Bot.* **80** 275–287 (*224*)

Niklas K J 1997b. *The evolutionary biology of plants* (Chicago: University of Chicago Press) (*130*)

Niklas K J and Spatz H-C 2004 Growth and hydraulics (not mechanical) constraints govern the scaling of tree height and mass; *Proc. Nat. Acad. Sci. USA* **101** 15661–15663 (*135*)

Nishikawa K C and Gans C 1996 Mechanisms of tongue protraction and narial closure in the marine toad *Bufo marinus. J. Exp. Biol.* **199** 2511–2529 (*28*)

Nobel P S 2005 *Physicochemical and environmental plant physiology, 3rd ed.* (Burlington MA: Elsevier) (*64, 76, 88, 89, 132, 134, 135, 165, 168, 173, 177, 194, 197, 205*)

Norberg U M 1990 *Vertebrate flight: mechanics, physiology, morphology, ecology and evolution* (Berlin: Springer-Verlag) (*111, 246*)

Norberg U M and Rayner J M V 1987 Ecological morphology and flight in bats (Mammalia: Chiroptera): wing adaptations, flight performance, foraging strategy and echolocation; *Phil. Trans. R. Soc. Lond.* **B316** 335–427 (*246*)

Nowak R M 1991 *Walker's mammals of the world, 5th ed.* (Baltimore: Johns Hopkins University Press) (*43, 120, 251, 257*)

Nowell A R M, Jumars P A and Fauchald K 1984 The foraging strategy of a subtidal and deep-sea deposit feeder; *Limnol. Oceanogr.* **29** 645–649 (*37*)

O'Brien K R, Ivey G N, Hamilton D P and White A M 2003 Simple mixing criteria for the growth of negatively buoyant phytoplankton; *Limnol. Oceanogr.* **48** 1326–1337 (*9*)

Okubo A and Levins S A 1989 A theoretical framework for data analysis of wind dispersal of seeds and pollen; *Ecology* 70 329–338 (*110*)

O'Leary D P and Ravasio M J 1994 Simulation of vestibular semicircular canal responses during righting maneuvers of a freely falling cat; *Biol. Cybernetics* 50 1–7 (*254*)

Ottaviani G and Tazzi A 1977 The lymphatic system; in *Biology of the reptilia, v6* (ed.) C Gans (London: Academic Press) pp 315–462 (*191*)

Pais A 1982 *Subtle is the lord—: the science and the life of Albert Einstein* (New York: Oxford University Press) (*60*)

Paley W 1802 *Natural theology* (London: Charles Knight 1856) (*224*)

Parker A R, Lawrence C R 2001 Water capture by a desert beetle; *Nature* 414 33–34 (*77*)

Parrott G C 1970 Aerodynamics of gliding flight of a black vulture *Coragyps atratus*; *J. Exp. Biol.* 53 363–374 (*110*)

Parry D A 1954 On the drinking of soil capillary water by spiders; *J. Exp. Biol.* 31 218–227 (*179*)

Parry D A and Brown R H J 1959 The jumping mechanism of salticid spiders; *J. Exp. Biol.* 36 654–664 (*28, 42, 49*)

Patek S N, Baio J E, Fisher B L and Suarez A V 2006 Multifunctionality and mechanical origins: ballistic jaw protrusion in trap-jaw ants; *Proc. Nat. Acad. Sci. US* 103 12787–12792 (*28, 42*)

Pedley T J, Brook B S and Seymour R S 1996 Blood pressure and flow rate in the giraffe jugular vein; *Philos. Trans. R. Soc. Lond.* B351 855–866 (*120*)

Pennak R W 1978 *Fresh-water invertebrates of the United States, 2nd ed.* (New York: John Wiley) (*155*)

Pennycuick C J 1960 Gliding flight of the fulmar petrel; *J. Exp. Biol.* 37 330–338 (*110*)

Pennycuick C J 1971 Gliding flight of the white-backed vulture, *Gyps africanus*; *J. Exp. Biol.* 55 13–38 (*110*)

Pennycuick C J 1975 Mechanics of flight; in *Avian biology, vol 5* (eds.) D S Farner and J R King (New York: Academic Press) pp 1–75 (*246*)

Pennycuick C J 1982 The flight of petrels and albatrosses (Procellariiformes), observed in South Georgia and its vicinity; *Phil. Trans. Roy. Soc.* B300 75–106 (*110*)

Pennycuick C J, Klaassen M, Kvist A and Lindström A 1996 Wingbeat frequency and the body drag anomaly: wind-tunnel observations on a thrush nightingale (*Luscinia luscinia*) and a teal (*Anas crecca*); *J. Exp. Biol.* 199 2757–2765 (*114*)

Peplowski M M and Marsh R L 1997 Work and power output in the hindlimb muscles of Cuban treefrogs *Osteopilus septentrionalis* during jumps; *J. Exp. Biol.* 200 2861–2870 (*48*)

Persson A 1998 How do we understand the Coriolis force? *Bull. Amer. Meteorol. Soc.* 79 1373–1385 (*97*)

Petroski H 1990 *The pencil: a history of design and circumstance* (New York: Knopf) (*233*)

Philpott J 1956 Blade tissue organization of foliage leaves of some Carolina shrub-bog species as compared with their Appalachian mountain affinities; *Bot. Gaz.* 118 88–105 (*90*)

Pickard, R S and Mill P J 1974 Ventilatory movements of the abdomen and branchial apparatus in dragonfly larvae (Odonata: Anisoptera); *J. Zool. Lond.* **174** 23–40 (*191*)

Pickard, W F 2003a The riddle of root pressure. I. Putting Maxwell's demon to rest; *Func. Pl. Biol.* **30** 121–134 (*194*)

Pickard, W F 2003b The riddle of root pressure. II. Root exudation at extreme osmolalities; *Func. Pl. Biol.* **30** 135–141 (*194*)

Pickard W F 2006 Absorption by a moving spherical organelle in a heterogeneous cytoplasm: implications for the role of trafficking in a symplast; *J. Theor. Biol.* **240** 288–301 (7)

Pittermann J and Sperry J S 2003 Tracheid diameter determines the extent of freeze-thaw induced cavitation in conifers; *Tree Physiol.* **23** 907–914 (*203*)

Pommen G D W and Craig D A 1995 Flow patterns around gills of pupal net-winged midges (Diptera: Blephariceridae): possible implications for respiration; *Can. J. Zool.* **73** 373–382 (*153, 154*)

Pringle A, Patek S N, Fischer M, Stotze J and Money N P 2005 The captured launch of a ballistospore; *Mycologia* **97** 866–871 (*33, 41, 50*)

Prosser C L, ed. 1973 *Comparative animal physiology, 3rd ed.* (Philadelphia: W B Saunders) (*175, 191*)

Purcell E M 1977 Life at low Reynolds number; *Am. J. Physics* **45** 3–11 (9)

Putz F E and Holbrook N M 1991 Biomechanical studies of vines; in *The biology of vines* (ed.) F E Putz and H A Mooney (Cambridge: Cambridge University Press) pp 73–97 (*221*)

Rabinowitz D and Rapp J K 1981 Dispersal abilities of seven sparse and common grasses from a Missouri prairie; *Amer. J. Bot.* **68** 616–624 (*110*)

Reilly D T and Burstein A H 1975 The elastic and ultimate properties of compact bone tissue; *J. Biomech.* **8** 393–405 (*221*)

Resh V H and Jamieson W 1988 Parasitism of the aquatic moth *Petrophila confusalis* (Lepidoptera: Pyralidae) by the aquatic wasp *Tanychela pilosa* (Hymenoptera: Ichneumonidae); *Entomol. News* **99** 185–188 (*153*)

Riisgård H U and Larsen P S 1995 Filter-feeding in marine macro-invertebrates: pump characteristics, modeling and energy cost; *Biol. Rev.* **70** 67–106 (*192, 203*)

Robinson D C E and Geils B W 2006 Modelling dwarf mistletoe at three scales: life history, ballistics and contagion; *Ecol. Model.* **199** 23–28 (30)

Rommel S A and Caplan H 2003 Vascular adaptations for heat conservation in the tail of Florida manatees (*Trichechas manatus latirostris*); *J. Anat.* **202** 343–353 (86)

Rommel S A, Pabst D A, McLellan W A and Potter C W 1992 Anatomical evidence for a countercurrent heat exchanger associated with dolphin testes; *Anat. Rec.* **232** 150–156 (86)

Rossini G A 1829 Guillaume Tell (Opera, especially Overture to) (*128*)

Rowe N P, Isnard S, Gallenmüller F and Speck T 2006 Diversity of mechanical architectures in climbing plants: an ecological perspective; in *Ecology and biomechanics* (eds.) A Herrel, T Speck and N P Rowe (Boca Raton FL: CRC Press) pp 35–59 (*220*)

Rubega M A and Obst B S 1993 Surface-tension feeding in phalaropes: discovery of a novel feeding mechanism; *Auk* 110 169–178 (*198*)

Sakai A and Larcher W 1997 *Frost survival of plants* (Berlin: Springer-Verlag) (*168*)

Sanderson S L, Cheer A Y, Goodrich J S, Graziano J D and Callan W T 1991 Crossflow filtration in suspension-feeding fishes; *Nature* 412 439–441 (*12*)

Sato K, Naito Y, Kato A, Niizuma Y, Watanuki Y, Charrassin J B, Bost C-A, Handrich Y and LeMaho Y 2002 Buoyancy and maximal diving depth in penguins: do they control inhaling air volume? *J. Exp. Biol.* 205 1189–1197 (*157*)

Schlesinger W H, Gray J T, Gill, D S and Mahall B E 1982 *Ceanothus megacarpus* chaparral: a synthesis of ecosystem processes during development and animal growth; *Bot. Rev.* 48 71–117 (*133*)

Schmidt-Nielsen B and Schmidt-Nielsen K 1950 Evaporative water loss in desert rodents in their natural habitat; *Ecology* 31 75–85 (*165*)

Schmidt-Nielsen K 1964 *Desert animals: physiological problems of heat and water* (Oxford: Oxford University Press) (*66, 88*)

Schmidt-Nielsen K 1972 *How animals work* (Cambridge: Cambridge University Press) (*87*)

Schmidt-Nielsen K 1981 Countercurrent systems in animals; *Sci. Amer.* 244 (5) 118–129 (*87*)

Schmidt-Nielsen K 1997 *Animal physiology: adaptation and environment, 5th ed.* (Cambridge: Cambridge University Press) (*16, 118, 157, 159, 193*)

Schmidt-Nielsen K, Schmidt-Nielsen B, Jarnum S A and Houpt T R 1957 Body temperature of the camel and its relation to water economy; *Amer. J. Physiol.* 188 103–112 (*88*)

Schmidt-Nielsen K, Taylor C R and Shkolnik A 1971 Desert Snails: problems of heat, water and food; *J Exp Biol.* 55 385–398 (*64*)

Schneider H, Wistuba N, Wagner H-J, Thürmer F and Zimmermann U 2000 Water rise kinetics in refilling xylem after desiccation in a resurrection plant; *New Phytol.* 148 221–238 (*198*)

Scholander P F 1955 Evolution of climatic adaptations in homeotherms; *Evolution* 9 15–26 (*86*)

Scholander P F 1957 The wonderful net; *Sci. Amer.* 196 (4) 96–107 (*86*)

Scholander P F, Claff C L and Sveinsson S L 1952 Respiratory studies of single cells. I. Methods; *Biol. Bull.* 102 157–177 (*162*)

Scholander P F, Flagg W, Hock R J and Irving L 1953 Studies on the physiology of frozen plants in the Arctic; *J. Cell. Comp. Physiol.* 42, **suppl 1** 156 (*168, 169, 171*)

Scholander P F, Hammel H T, Bradstreet E D and Hemmingsen EA 1965 Sap pressure in vascular plants; *Science* 148 339–346 (*133*)

Scholander P F and Krog J 1957 Countercurrent heat exchange and vascular bundles in sloths; *J. Appl. Physiol.* 10 405–411 (*86*)

Scholander P F and Maggert J E 1971 Supercooling and ice propagation in blood from Arctic fishes; *Cryobiology* 8 371–374 (*166*)

Scholander P F and Schevill W E 1955 Counter-current vascular heat exchange in the fins of whales; *J. Appl. Physiol.* 8 279–282 (*86*)

Scholander P F and Van Dam C L 1954 Secretion of gases against high pressures in the swimbladder of deep sea fishes; *Biol. Bull.* **107** 247–259 (*86*)

Scholander P F, Walters V, Hock R and Irving L 1950 Heat regulation in some arctic and tropical mammals and birds; *Biol. Bull.* **99** 237–258 (*81*)

Schuetz D and Taborsky M 2003 Adaptations to an aquatic life may be responsible for the reversed sexual size dimorphism in the water spider, *Argyroneta aquatica*; *Ecol. Evol. Res.* **5** 105–117 (*144*)

Schulz J R, Norton A G and Gilly W F 2004 The projectile tooth of a fish-hunting snail: *Conus catus* injects venom into fish prey using a high-speed ballistic mechanism; *Biol. Bull.* **207** 77–79 (*191*)

Schumacher G J and Whitford H A 1965 Respiration and P^{32} uptake in various species of freshwater algae as affected by a current; *J. Phycol.* **1** 78–80 (*10*)

Schwartz-Dabney C L and Dechow P C 2003 Variations in cortical material properties throughout the human dentate mandible; *Am. J. Phys. Anthropol.* **120** 252–277 (*222*)

Sewell A 1877 *Black Beauty* (London: Jarrold and Sons) (*239*)

Seymour R S and Arndt J O 2004 Independent effects of heart-head distance and caudal pooling on pressure regulation in aquatic and terrestrial snakes; *J. Exp. Biol.* **207** 1305–1311 (*121*)

Seymour R S and Blaylock A J 2000 The principle of Laplace and scaling of ventricular wall stress and blood pressure in mammals and birds; *Physiol. Biochem. Zool.* **73** 389–405 (*117, 120*)

Shadwick R E 1994 Mechanical organization of the mantle and circulatory system of cephalopods; *Mar. Fresh. Behav. Physiol.* **25** 69–85 (*203*)

Sherrington C (1906) *The integrative action of the nervous system* (New Haven, CT: Yale University Press; 2nd ed. 1947) (*235*)

Shimek R L and Kohn A J 1981 Functional morphology and evolution of the toxoglossan radula; *Malacologia* **20** 423–438 (*37*)

Simpson G G 1970 The Argyrolagidae, extinct South American marsupials; *Bull. Mus. Comp. Zool.* **139** 1–86 (*257*)

Sláma K 2003 Mechanical aspects of heartbeat reversal in pupae of *Manduca sexta*; *J. Insect. Physiol.* **49** 645–657 (*206*)

Smith A P 1972 Buttressing of tropical trees: a descriptive model and new hypotheses; *Am. Nat.* **106** 32–46 (*137*)

Socha J J 2002 Gliding flight in the paradise tree snake; *Nature* **418** 603–604 (*111*)

Socha J J, O'Dempsey T and LaBarbera, M 2005. A three-dimensional kinematic analysis of gliding in a flying snake, *Chrysopelea paradisi*; *J. Exp. Biol.* **208** 1817–1833 (*111, 256*)

Solari C A, Ganguly S, Kessler J O, Michod R E and Goldstein R E 2006 Multicellularity and the functional interdependence of motility and molecular transport; *Proc. Nat. Acad. Sci. US* **103** 1353–1358 (*2, 7*)

Solga A, Cerman Z, Striffler B F, Spaeth M and Barthlott W 2007 The dream of staying clean: lotus and biomimetic surfaces; *Bioinsp. Biomim.* **2** S126–S134 (*147*)

Soppela P, Neimenen M and Saarela S 1992 Water intake and its thermal energy cost in reindeer fed lichen or various protein rations during winter; *Acta Physiol. Scand.* **145** 65–73 (*175*)

Southwick E E and Moritz R F A 1987 Social control of air ventilation in colonies of honey bees, *Apis mellifera*; *J. Insect Physiol.* **33** 623–626 (*196*)

Sparks J P, Campbell G S and Black A R 2001 Water content, hydraulic conductivity, and ice formation in winter stems of *Pinus contorta*: a TDR case study; *Oecologia* **127** 1432–1439 (*169*)

Srinivasan M and Ruina A 2006 Computer optimization of a minimal biped model discovers walking and running; *Nature* **439** 72–75 (*126*)

Stamp N E and Lucas J R 1983 Ecological correlates of explosive seed dispersal; *Oecologia (Berlin)* **59** 272–278 (*30, 42*)

Stamp N E and Lucas J R 1990 Spatial patterns and dispersal distances of explosively dispersing plants in Florida sandhill vegetation; *J. Ecol.* **78** 589–600 (*30, 41*)

Steffensen J F 1985 The transition between branchial pumping and ram ventilation in fishes: energetic consequences and dependence on water oxygen tension; *J. Exp. Biol.* **114** 141–150 (*199*)

Stevens E D and Lightfoot E N 1986 Hydrodynamics of water flow in front of and through the gills of skipjack tuna; *Comp. Biochem. Physiol.* **83A** 255–259 (*11, 203*)

Stewart I (2004) Quantizing the classical cat; *Nature* **430** 731–732 (*253*)

Stokes V J, Morecroft M D, Morison J I L 2006 Boundary layer conductance for contrasting leaf shapes in a deciduous broadleaves forest canopy; *Agr. For. Meteorol.* **139** 40–54 (*67*)

Storey K B 2006 Reptile freeze tolerance: metabolism and gene expression; *Cryobiology* **52** 1–16 (*168*)

Storey K B and Storey J M 1992 Natural freeze tolerance in ectothermic vertebrates; *Annu. Rev. Physiol.* **54** 619–637 (*168*)

Stride G O 1955 On the respiration of an aquatic African beetle, *Potamodytes tuberosus* Hinton; *Ann. Entomol. Soc. Am.* **48** 344–351 (*145, 151, 152*)

Suzuki K and Taniguchi Y 1972 Effect of pressure on biopolymers and model systems; *Symp. Soc. Exp. Biol.* **26** 103–124 (*142*)

Swaine M D and Beer T 1977 Explosive seed dispersal in *Hura crepitans* L. (Euphorbiaceae); *New Phytol.* **78** 695–708 (*29*)

Swaine M D, Dakubu T and Beer T 1979 On the theory of explosively dispersed seeds: a correction; *New Phytol.* **82** 777–781 (*29, 43*)

Taylor F W 2005 *Elementary climate physics* (Oxford: Oxford University Press) (*115*)

Taylor G K and Krapp H G 2008 Sensory systems and flight stability: what do insects measure and why? in *Advances in insect physiology, vol 34* (eds.) J Casas and S J Simpson (London: Elsevier) pp 231–316 (*249*)

Taylor, M A 1993 Stomach stones for feeding or buoyancy? The occurrence and function of gastroliths in marine tetrapods; *Phil. Trans. R. Soc. Lond.* **B341** 163–175 (*155*)

Taylor P E, Card G, House J, Dickinson M H and Flagan R C 2006 High-speed pollen release in the white mulberry tree, *Morus alba* L; *Sex. Plant Reprod.* **19** 19–24 (*33, 40, 41*)

Tennekes H 1996 *The simple science of flight: from insects to jumbo jets* (Cambridge, MA: MIT Press) (*110, 113*)

Thom A and Swart P 1940 The forces on an aerofoil at very low speeds; *J. Roy. Aero. Soc.* 44 761–770 (*109*)

Thompson D'A W 1942 *On growth and form* (Cambridge: Cambridge University Press) (*39*)

Thompson M V and Holbrook N M 2003 Scaling phloem transport: water potential equilibrium and osmoregulatory flow; *Plant Cell Environ.* 26 1561–1577 (*194*)

Thorpe W H 1950 Plastron respiration in aquatic insects; *Biol. Rev.* 25: 344–390 (*145*)

Thorpe W H and Crisp D J 1947 Studies on plastron respiration. I. The biology of *Aphelocheirus* [Hemiptera, Aphelocheiridae (Naucoridae)] and the mechanism of plastron retention; *J. Exp. Biol.* 24 227–269 (*145, 146, 147*)

Ting L H, Blickhan R and Full R J 1994 Dynamic and static stability in hexapedal runners; *J. Exp. Biol.* 197 251–269 (*241*)

Toro E, Herrel A, Vanhooydonck B and Irschick D J 2003 A biomechanical analysis of intra- and interspecific scaling of jumping and morphology in Caribbean lizards; *J. Exp. Biol.* 206 2641–2652 (*43*)

Trager G C, Hwang J-S and Strickler J R 1990 Barnacle suspension feeding in variable flow; *Mar. Biol.* 105 117–128 (*203*)

Trail F, Gaffoor I and S Vogel 2005 Ejection mechanics and trajectory of the ascospores of *Gibberella zeae* (anamorph *Fusarium graminearum*); *Fungal Genet. Biol.* 42 528–533 (*32, 41, 110, 193, 203*)

Tucker V A 2000 Gliding flight: drag and torque of a hawk and a falcon with straight and turned heads, and a lower value for the parasite drag coefficient; J. Exp. Biol. 203 3733–3744 (*114*)

Tucker V A, Cade T J and Tucker A 1998 Diving speeds and angles of a gyrfalcon (*Falco rusticoles*); *J. Exp. Biol.* 201 2061–2070 (*30*)

Tucker V A and Heine C 1990 Aerodynamics of gliding flight in a Harris' hawk, *Parabuteo unicinctus*; *J. Exp. Biol.* 149 469–489 (*110*)

Tucker V A and Parrott, G C 1970 Aerodynamics of gliding flight in a falcon and other birds; *J. Exp. Biol.* 52 345–367 (*108, 110*)

Turner J S 1987 The cardiovascular control of heat exchange: consequences of body size; *Am. Zool.* 27 69–79 (*74*)

Turner J S 1988 Body size and thermal energetics: how should thermal conductance scale? *J. Thermal Biol.* 13 103–117 (*74*)

Turner J S 2000 *The extended organism: the physiology of animal-built structures* (Cambridge, MA: Harvard University Press) (*199*)

Turner J S and Picker M D 1993 Thermal ecology of an embedded dwarf succulent from southern Africa (*Lithops* spp: Mesembryanthemaceae); *J. Arid Environ.* 24 361–385 (*88*)

Tyree M T 2001 Capillarity and sap ascent in a resurrection plant: does theory fit the facts? *New Phytol* 150 9–11 (*198*)

Tyree M T and Zimmermann M H 2002 *Xylem structure and the ascent of sap* (Berlin: Springer-Verlag) (*194*)

Usherwood J R and Bertram J E H 2003 Understanding brachiation: insight from a collisional perspective; *J. Exp. Biol.* 206 1631–1642 (*28*)

Vallee R B 1998 *Molecular motors and the cytoskeleton* (San Diego: Academic Press) (6)

van Bel A 1993 Strategies of phloem loading; *Annu. Rev. Pl. Physiol. Pl. Mol. Biol.* **44** 253–281 (194)

Verkaar H J, Schenkeveld A J and van de Klashorst M P 1983 The ecology of short-lived forbs in chalk grasslands: dispersal of seeds; *New Phytol.* **95** 335–344 (110)

Vincenti W G 1990 *What engineers know and how they know it: analytical studies from aeronautical history* (Baltimore: Johns Hopkins University Press) (244)

Vogel S 1966 Flight in *Drosophila*. I. Flight performance of tethered flies; *J. Exp. Biol.* **44** 567–578 (114)

Vogel S 1968 "Sun leaves" and "shade leaves": differences in convective heat dissipation; *Ecology* **49** 1203–1204 (81)

Vogel, S 1976 Flows in organisms induced by movements of the external medium in *Scale effects in animal locomotion* (ed.) T J Pedley (London: Academic Press, Ltd) pp 285–97 (180)

Vogel S 1978 Evidence for one-way valves in the water-flow system of sponges; *J. Exp. Biol.* **76** 137–148 (203)

Vogel S 1984 The thermal conductivity of leaves; *Can. J. Bot.* **62** 741–744 (67)

Vogel S 1988 *Life's devices: the physical world of animals and plants* (Princeton, NJ: Princeton University Press) (22)

Vogel S 1989 Drag and reconfiguration of broad leaves in high winds; *J. Exp. Bot.* **40** 941–948 (136, 138, 222)

Vogel S 1992 Twist-to-bend ratios and cross-sectional shapes of petioles and stems; *J. Exp. Bot.* **43** 1527–1532 (230)

Vogel S 1994a Dealing honestly with diffusion; *Amer. Biol. Teacher* **56** 405–407 (3)

Vogel S 1994b *Life in moving fluids* (Princeton, NJ: Princeton University Press) (8, 12, 101, 109, 114, 150, 153, 198, 202)

Vogel S 1995a. Pressure versus flow in biological pumps; in *Biological fluid dynamics: symposium of the society for experimental biology* (eds.) C P Ellington and T J Pedley (Cambridge: The Company of Biologists, Ltd.) pp 297–304 (184, 201)

Vogel S 1995b. Twist-to-bend ratios of woody structures; *J. Exp. Bot.* **46** 981–985 (219, 220, 221)

Vogel S 1996a. Blowing in the wind: storm-resisting features of the design of trees; *J. Arboriculture* **22** 92–98 (136)

Vogel S 1996b. Diversity and convergence in the study of organismal function; *Israel J. Zool.* **42** 297–305 (223)

Vogel S 1998 Exposing life's limits with dimensionless numbers; *Phys. Today* **51**(11) 22–27 (16)

Vogel S 2001 *Prime mover: a natural history of muscle* (New York: W. W. Norton) (48)

Vogel S 2003 *Comparative biomechanics: life's physical world* (Princeton, NJ: Princeton University Press) (53, 55, 125, 130, 241)

Wagner H 1925 Über die Entstehung des dynamischen Auftriebes von Tragflügeln; *Z. Angew. Math. Mech.* **5** 17–35 (*105*)

Wagner T, Neinhuis C and Barthlott W 1996 Wettability and contaminability of insect wings as a function of their surface sculpture; *Acta Zool.* **77** 213–225 (*147*)

Wainwright S A 1988 *Axis and circumference* (Cambridge, MA: Harvard University Press) (*222*)

Wainwright S A, Biggs W D, Currey J D and Gosline J M 1976 *Mechanical design in organisms* (New York: John Wiley) (*155*)

Wallace H K 1942 A revision of the burrowing spiders of the genus *Geolycosa* (Areneae, Lycosidae); *Am. Midl. Nat.* **27** 1–62 (*179*)

Ward-Smith A J 1984 *Biophysical aerodynamics and the natural environment* (Chichester: Wiley-Interscience) (*110*)

Warnaars T A and Hondzo M 2006 Small-scale fluid motion mediates growth an nutrient uptake of *Selenastrum capricornutum; Freshwater Biol.* **51** 999–1015 (*14*)

Wauthy G, Leponce M, Banaï N, Sylin G and Lions J-C 1998 The backward jump of a box moss mite; *Proc. R. Soc. Lond.* B **265** 2235–2242 (*28, 41*)

Weast R C (ed.) 1987 *CRC handbook of chemistry and physics, 68th ed.* (Boca Raton: CRC Press) (*98*)

Webb P W 2002 Control of posture, depth, and swimming trajectories of fishes; *Integ. Comp. Biol.* **42** 94–101 (*250*)

Webber D M, Aitkin J P and O'Dor R K 2000 Costs of locomotion and vertical dynamics of cephalopods and fish; *Physiol. Biochem. Zool.* **73** 651 (*156*)

Weis-Fogh T 1956 Biology and physics of locust flight. II. Flight performance of the desert locust; *Phil. Trans. Roy. Soc. Lond.* B**239** 459–510 (*114*)

Weis-Fogh T 1961 Thermodynamic properties of resilin, a rubber-like protein; *J. Molec. Biol.* **3** 648–667 (*59*)

Weiss M R 2003 Good housekeeping: why do shelter-dwelling caterpillars fling their frass? *Ecol. Letts.* **6** 361–370 (*28*)

Wells M J 1987 The performance of the octopus circulatory system: a triumph of engineering over design; *Experientia* **43** 487–499 (*203*)

Werner P A and Platt W J 1976 Ecological relationships of co-occurring goldenrods (*Solidago*: Compositae); *Amer. Natur.* **110** 959–971 (*110*)

Westlake D R 1977 Some effects of low-velocity currents on the metabolism of aquatic macrophytes; *J. Exp. Bot.* **18** 187–205 (*10*)

Wharton D A 2002 *Life at the limits: organisms in extreme environments* (Cambridge: Cambridge University Press) (*167, 182*)

Whitaker D L, Webster L A and Edwards J 2007 The biomechanics of *Cornus canadensis* stamens ar ideal for catapulting pollen vertically; *Func. Ecol.* **21** 219–225 (*41*)

White F M 1974 *Viscous fluid flow* (New York: McGraw-Hill) (*22*)

Wigglesworth V B 1963 Origin of wings in insects; *Nature* **197** 97–98 (*112*)

Wilga C D and Lauder G V 2002 Function of the heterocercal tail in sharks: quantitative wake dynamics during steady horizontal swimming and vertical maneuvering; *J. Exp. Biol.* **205** 2365–2374 (*156*)

Willmer P, Stone G and Johnston I 2000 *Environmental physiology of organisms* (Oxford: Blackwell Science Ltd.) (*167, 176, 181, 182*)

Withers P C 1981 An aerodynamic analysis of bird wings as fixed aerofoils; *J. Exp. Biol.* **90** 143–162 (*109*)

Witztum A and Schulgasser K 1995a The mechanics of seed expulsion in Acanthaceae; *J. Theor. Biol.* **176** 531–542 (*30, 42, 52*)

Witztum A and Schulgasser K 1995b Seed dispersal ballistics in *Blepharis ciliaris*; *Israel J. Pl. Sci.* **43** 147–150 (*42*)

Woodson C B, Webster D R, Weissburg M J and Yen J 2007 Cue hierarchy and foraging in calanoid copepods: ecological implications of oceanographic structure; *Mar. Ecol. Prog. Ser.* **330** 163–177 (*14*)

Woodward B L, Winn J P and Fish F E 2006 Morphological specializations of baleen whales associated with hydrodynamic performance and ecological niche; *J. Morphol.* **267** 1284–1294 (*251*)

Yanoviak S P, Dudley R and Kaspari M 2005 Directed aerial descent in canopy ants; *Nature* **433** 624–626 (*109, 252*)

Yarwood C E and Hazen W E 1942 Vertical orientation of powdery mildew conidia during fall; *Science* **96** 316–317 (*110*)

Yigit N, Güven T, Bayram A and Cavusoglu K 2004 A morphological study on the venom apparatus of the spider *Agelena labyrinthica* (Araneae, Agelenidae); *Turk. J. Zool.* **28** 149–153 (*191*)

Yokoyama, T 2004 Motor or sensor: a new aspect of primary cilia function; *Anat. Sci. Int.* **79** 47–54 (*193*)

Yom-Tov Y 1971 Body temperature and light reflectance in two desert snails; *Proc. Malacol. Soc. Lond.* **39** 319–326 (*64*)

Young B A, Dunlap K, Koenig K and Singer M 2004 The buccal buckle: the functional morphology of venom spitting in cobras; *J. Exp. Biol.* **207** 1383–1394 (*191*)

Young B A, Phelan M, Morain M, Ommundsen M and Kurt R 2003 Venom injection by rattlesnakes (*Crotalus atrox*): peripheral resistance and the pressure balance hypothesis; *Can. J. Zool.* **81** 313–320 (*191, 203*)

Zeuthen E 1943 A Cartesian diver micro respirometer with a gas volume of 0.1 µl; *C. R. Trav. Lab. Carlsberg, ser. chim.* **24** 479–518 (*162*)

Zimmermann M H 1971 Transport in the xylem; in *Trees: structure and function* (ed.) M H Zimmermann and C L Brown (New York: Springer-Verlag) pp 169–220 (*203*)

Zimmermann M H 1983 *Xylem structure and the ascent of sap* (Berlin: Springer-Verlag) (*134*)

Zweifach B W 1974 Quantitative studies of microcirculatory structure and function. I. Analysis of pressure distribution in the terminal vascular bed in cat mesentery; *Circulation Res.* **34** 843–857 (*117*)

Index

Italic entries refer to figures, "p" entries to plates.

absorptivity. *See* emissivity
acceleration: cost, 123; energy storage
 and, 49; gravitational, 95, 101; limit by
 stress, 54–56; mass vs., 95, 116;
 non-projectiles, 57; power amplification,
 47–49; projectiles, *44*; scaling, 40–45,
 54–56; tabulated values, 41; to terminal
 velocity, 102. *See also specific
 organisms*
acceleration reaction, 104
Acer rubrum, 223
Acris, jump, 43
adaptation. *See* evolution
added mass coefficient, 105
advance ratio, 114, 260
aeroelasticity, 227
air: compressibility, 54, 99; energy storage,
 53, 54; mean free path, 104; sinking
 rates in, 101; specific heat, 66; "still air,"
 89; thermal conductivity, 68, 81;
 thermal expansion coefficient, 70; under
 water, 141, 143–54; water content, 60,
 79; within soil, 165
airfoils, 106–12, *109*
Albizzia leaf orientations, 65, p4
Alexander, R. McNeill, 123
algae, p1; infrared reflection, 64; Péclet
 number, 7, 13. *See also* phytoplankton
Alsomitra, 111, 244, 245
Ammospermophilus, cooling, 68
Amphicteis, fecal trajectory, 37
anhingas, 158, p8
anisotropy, 228
Anolis jump, 43
Antechinomys, 257
antelope, acceleration, 40
antelope ground squirrel, 68
antifreezes. *See* cryoprotectants
ants: gliding, 109, 252; trap-jaw acceleration, 57; trap-jaw jump, 28, 42
aorta, pressure in, 117
Aphelocheirus, 146

Arceuthobium seed: acceleration, 44; seed
 launch, 56
Archimedean screw, *190*
area, second moment of, *215*
Arenicola, 196
Argentavis, 107, 246
Argyroneta aquatica, 144
Arrhenius equation, 59
Artemia salina, 182
arthropods: ciliary lack, 207; water
 condensation, 77, 176, 177. *See also*
 insects
ascidians: heart pumps, 206; suspension
 feeding, 199
Ascobolus spore: acceleration, 41;
 trajectory, 25, 32
aspect ratio, 245
Aulocodes, 153
Auricularia spore: acceleration, 41;
 launching, 50; trajectory, 25, 33
autogyration, 106, 244
Avogadro's number, 16, 60, 260

bacteria: swimming speed, 9; temperature
 tolerance, 58
Bagnold number, 101, 259
ballistics, 18–38, 39–57; drag, 19; energy
 supply, 50; launching devices, 51; lift
 role, 33; trajectories, 18–38, 19;
 underwater, 37. *See also* trajectories
bamboo, 136, 219, 234
banana petioles, 224
basilisk lizard, 98
basketball, 26, 27
Bassett term, 105
bat flight, 246
Baudoin number, 98, 144, 259
Bauhinia, 218, 220
beams: cantilever, 131; grooved, p11;
 petioles as, 222; stresses within, 210,
 211, 212
bee hive ventilation, 196